Comic Book Geographies
Edited by Jason Dittmer

MEDIA GEOGRAPHY AT MAINZ

Edited by
Anton Escher
Chris Lukinbeal
Stefan Zimmermann

Managing Editor
Christina Kerz

Volume 4

Comic Book Geographies

Edited by Jason Dittmer

Franz Steiner Verlag

Druck unterstützt durch das Zentrum
für Interkulturelle Studien

ZIS Zentrum für Interkulturelle Studien MAINZ

Bibliographische Information der Deutschen Bibliothek
Die Deutsche Bibliothek verzeichnet diese Publikation in der Deutschen Nationalbibliographie; detaillierte bibliographische Daten sind im Internet über <http://dnb.ddb.de> abrufbar.

ISBN 978-3-515-10269-8

Gedruckt auf säurefreiem, alterungsbeständigem Papier.
Druck: Laupp und Göbel, Nehren
Printed in Germany

For my father, Kenneth Dittmer (1927–2012)

Geographers usually see themselves in a tradition of Alexander von Humboldt and we would like to highlight his position that everything in our world is interdependent. Deriving from that assumption geography has to be seen as an unlimited science, a discipline that knows its borders but has none.

Anton Escher, Chris Lukinbeal and Stefan Zimmermann,
Mainz, January 2008

CONTENTS

Daniel Merlin Goodbrey

FOREWORD

I'm ten years old and I'm about to make my first comic.

The comic is going to be one page long and once it's done I'm going to photocopy it and sell it in my parents' antique shop. When I make my second comic it's going to be two pages long and I'm going to figure out how to make the photocopier print onto both sides of the paper. When I make my third comic it's going to be a proper little booklet: eight pages long, constructed out of two pages of folded A4 and stapled in the middle. It's going to be part of a series of six or so issues about a talking sausage and an egg who fight against an evil frying pan. But that's not really important right now.

This is the first comic. I have a pencil, a ruler, a black pen and a blank sheet of paper. A blank space. I take the pencil and the ruler and I divide up the blank space into nine smaller blank spaces of exact and equal size. Happy with my handiwork, I take the pen and the ruler and make the borders of each space permanent. I pause. Nine blank panels await me, each measure of page space carefully outlined in black.

I pick up the pencil and I start to draw.

* * *

It's ten years later and I don't make comics anymore.

It was the drawings that were the problem. I never could make the pictures on the paper match up satisfactorily to the ones in my head. I'm in the first year of my degree in multimedia design and I'm learning bits and pieces of 3D modelling, image manipulation and animation. I'm writing collaborative fiction online and picking up the basics of coding but I'm still not making comics. Then at some point something sets me to wondering. Could I use a computer to create the artwork for a comic? In my second semester I read a book called *Understanding Comics* by Scott McCloud. Something in my brain goes click.

Two years pass. I finish my undergraduate studies and start working towards a Master's degree in interactive narrative. Somewhere along the line I also start dabbling in digital comics. I use a mix of 3D and photo manipulation to solve my drawing problem and create a handful of little experimental pieces. Most significantly I provide the artwork for an early webcomic series called *Rust*. *Rust* does some interesting things with its layout, incorporates pop-ups and animation as storytelling tools and gets appreciative nods in the right places. It pushes me to develop my digital art skills further and causes me to consider what else might be possible for comics on the web.

On my Master's programme I'm learning about multicursal stories with branching pathways and ergodic narratives that change each time they're read. I start to think about what it would be like to apply these ideas in my webcomics. I read another book by Scott McCloud called *Reinventing Comics*. There's a chapter in the book entitled "The Infinite Canvas" in which McCloud makes his case for the unrealised potential of digital comics. I look around online and wonder why more people aren't exploring the ideas McCloud outlined in that chapter. Something else in my brain goes click. This time it's a louder click, followed by a series of increasingly impressive, pixelated explosions.

* * *

It's twelve years later and I'm writing the foreword to this book.

I'm sitting down at my computer and I'm thinking about space. I'm thinking about the last twelve years of my work in comics and how I've used space and made spaces. I'm thinking about how that use of space has changed between the page and the screen and the gallery wall. I'm thinking about interconnections between space and time; between story and shape; between word and image. And then I'm thinking further back. To that first little comic I made when I was ten. That first division of space into nine little blank boxes. They could have contained anything, those boxes. The potential of everything that followed was encoded right there in the space of that very first page.

Nine little blank boxes. It seems as good a place to start as any.

* * *

In traditional printed comic books, space is a finite resource. Every event depicted within the comic has to be allotted a share of that resource. The total space that constitutes the comic's narrative is divided into panels of various sizes and those panels are organised around fixed, homogenised groupings that we call pages. Stories are told across fixed, pre-determined page counts. For a comic creator working within the confines of print, space immediately becomes a premium commodity. Accordingly, creators quickly learn to get the most narrative impact possible out of every page and every panel available to them.

In my own work for print I tend to favour simplicity in terms of page layout. I start by writing dialogue and sometimes short stage directions, then break that text down into panels and pages to get a feel for how the story might lay out. When it comes to arranging the panels on the page, I often use compositions consisting solely of page-wide panels of landscape proportions. These are easy for the reader to follow and give a nice smooth rhythm to the action on the page. Their dimensions also recall that of the cinema screen and allow for similar compositional tricks in the artwork. For more conversational or condensed narratives I like to work a variation of a grid, most often either six panels or a classic nine panel grid.

If I'm writing for another artist to illustrate I try not to give too much direction at all on page layout, preferring for the artist to decide themselves on the best way to tackle the space of the page. Although even then, specifying the number of panels in each page of script has a big influence on the story. Three or four panels and you've got a fast, pacey read. Cram in twelve or sixteen and suddenly you've got room for a lot of plot-per-page but you're likely to wear your artist out pretty quickly. A good sweet spot for commercial work seems to be somewhere around six to eight panels; enough panels to carry a good chunk of narrative on each page but still plenty of room for the artist to have some fun.

In contrast to creating a comic for print, creating a digital comic means operating in a medium where space is neither fixed nor finite. Instead, space becomes plastic. It becomes slippery; it becomes mutable; it becomes infinitely configurable and reconfigurable. Gone are homogenised groupings of panels and fixed page counts. Gone is the constant nagging inner voice telling you to cram as much content as you can into the smallest amount of space possible. Hell, if you want them to be, gone are pages altogether.

And even if you decide to keep them around, your relationship with the page in the digital world is changed. Of course there's the obvious difference in dimensions; comics native to the screen tend to be in landscape layout rather than portrait. Beyond that though, there lies a more fundamental change; put simply, you just can't trust pages on the screen. They change. The panels swap about or turn up one at a time or slide around without warning. Images that used to have the decency to keep still can suddenly move when you're not looking at them. Worse, they can even move when you are looking at them. A page can be different every time you read it. Only the page and the page's creator would know; sitting together somewhere in the dark, laughing at you behind your back.

With such potential for spatial and narrative trickery, it's no wonder I fell in love with digital comics. The first bloom of that love was the hypercomic *Sixgun*, which I created as the major project for my Master's degree in 2001. In essence, a hypercomic is a comic with a multicursal narrative structure. Cursality is an idea that relates to the concept of the maze and the realisation that multiple different and divergent pathways might exist within a maze. A unicursal maze has only one path, no matter how convoluted, while a multicursal maze has many different possible pathways to explore and navigate. Multiple paths through a comic narrative mean that the reader is faced by repeated choices that can have a direct impact on the shape and outcome of the story they are reading. In a hypercomic, these choices may typically determine either the sequence in which events are encountered, the outcome of events or the point of view through which events are seen.

It wouldn't be too unfair to describe the rationale behind *Sixgun* as a desire to throw lots of different ideas about hypercomics at the wall and see what sticks. But at the core of the piece I was trying to resolve an essential conflict about the use of space. Comics is a spatial medium; all the elements in a comic exist in a fixed spatial relationship. In a hypertext like the World Wide Web, pages are linked together via hyperlinks. The only spatial relationship between pages is the one inferred by the hyperlinks, which typically do not provide a fixed visual element to tie pages together spatially. So the question became how to best preserve the fixed spatiality of comics within the less spatially-fixed hypertext medium.

In a typical hypertext a hyperlink might be a single word or image within the larger body of text. The solution I found for comics was to always treat whole panels as hyperlinks rather than any smaller component. The contents and position of the panel acting as the hyperlink would then be kept the same across both of the pages being linked together. When the panel is clicked the reader is linked from one page to another but, while the panels surrounding the hyperlinked panel change, the hyperlinked panel itself is unchanged. In this way the hyperlinked panel becomes the spatial constant tying the comic's multicursal network together.

Somewhere in the planning process of the stories that made use of these hyperlinked panels, I began to draw out flow diagrams to show how all the pages linked together. For some reason I found these diagrams fascinating. Fascinating enough that I started to insert them into the artwork of the comic itself. *Sixgun's* protagonists would occasionally encounter them – as a mural on a wall or in the design of a tattoo – and speculate obliviously as to the nature of what they were seeing.

In *Reinventing Comics* McCloud proposed the concept of the 'Infinite Canvas' – in essence the idea of treating the screen not as a page but as a window onto much larger configurations of panels. As part of this idea he proposed that the shape of an entire comic's layout could be directly linked to its narrative or thematic content. I became increasingly interested in figuring out how to make this idea work for a hypercomic. I began work on a new piece called *Doodleflak* that consisted of a series of branching pathways of panels configured into the shape of a snowflake. The shape of the whole flake would be visible at the beginning of the comic and then clicking on one section would zoom the canvas to focus on a single section of the narrative, with even smaller branches available at a further level of zoom.

It was the addition of this ability to zoom that really started to change the nature of space in my comics. The reader didn't just start with the overall shape of the narrative and then zoom in to read the comic; at any point they could zoom back out to see how the current position within the narrative related to the shape of the story as a whole. The zoom added a level of continuity to the reading experience and opened up ideas of relative scale as a new storytelling tool. I continued to explore the possibilities of the form for the next few years, across a whole series of zooming *Infinite Canvas* hypercomics.

Sometimes creating a new comic meant finding a shape for the story and other times it meant finding a story in the shape. So with some hypercomics I started with a story or theme I wanted to explore and let the final shape of the comic evolve alongside the development of the narrative. But with others I began by laying down a certain structure or shape to experiment with and then let that specific configuration of space drive the creation of the story. I've always found a few self-imposed restrictions to be a useful drive for creativity. Focusing first on the underlying spatial or interactive structure of a piece has often felt like the natural place to start. Here the real advantage of digital space is not in the leaving behind of all limitations but the ability to choose and configure your own limits to best suit the work in question.

Another advantage of the screen is that space ceases to be a carefully rationed commodity. With space no longer at a premium, the potential to experiment with the spatial relationship between panels becomes a much more appealing prospect. Larger gaps between panels can suggest a longer passage of time between each moment. Stretched out panels or the same image repeated across multiple panels can suggest time being dragged out and slowed down. Without the constant of the page turn, you're free to establish your own panel groupings and your own rhythm of advancement through the story. Even better, that rhythm doesn't have to remain fixed for the entire story but can instead be adjusted to underscore key shifts or transitions within the narrative.

As I built a reputation for experimental comics, opportunities arose that pushed my work in new directions. One such new direction saw me shift my exploration of hypercomics from digital to physical space. This shift began with my involvement in the Potential Comics project at the Institute of Contemporary Art in London. The original *PoCom* wall comic to which I contributed was a collaboration between eighteen cartoonists that measured seventeen meters long and two meters tall. This hypercomic was designed specifically to work with the flow of the space in which it was displayed, running from right to left rather than the traditional left to right in order to match the flow of foot traffic in the gallery.

The reversal of reading direction may seem like a small change, but for me it pointed the way towards a whole new avenue of enquiry. I began to wonder what comics might become once they were totally freed from the constraints and legacy of the printed page. What new spatial configurations might be possible? How could you create comics that really used and became part of the physical space in which they were displayed? I got the chance to explore these ideas further with the *Casita Situations* – a collaboration between myself, architect Valerio Ferrari and cartoonist David Baillie. Together we created a customisable hypercomic installation for the walls of a children's mental health clinic in Biobigny, Paris.

The *Casita Situations* consisted of a series of large panels featuring a recurring cast of characters that the staff could arrange in various different configurations around the clinic. The environments depicted in the panels began as extensions of the interior of the clinic but then opened out to reveal fantastical situations and locations beyond the walls of the building. The children would be left to create their own stories out of the characters and situations on show, influenced by the different panel arrangements chosen each week by the staff. This use of comics to extend the spaces of the building out into the territory of the imagination was an idea I found fascinating. I was able to explore the idea further in my next major gallery piece, *The Archivist.*

The Archivist was an installation that took up one floor of the Pumphouse Gallery in London as part of the *Hypercomics* exhibition in 2010. The piece documented the life of a lone archivist working in an alternate reality where the Pumphouse had become an archive to the life of a fictional rock star. The space of the building itself and the surrounding grounds featured heavily in the piece, shifting in appearance as the line between reality and videogames blurred together in the protagonist's dreams. The comic was divided across three walls of the floor while a fourth, virtual wall was represented by the printing of a web address on the room's main window. This web address moved the hypercomic back into digital space, where the story grew and expanded online over the course of the exhibition.

In trying to design a comic that worked both in the real space of the gallery and the digital space of the web, *The Archivist* brought me back full circle to examine the medium of the screen with fresh eyes. It started me thinking about videogames and comics and how both, on some fundamental level, were about explorations of space. How might the two media hybridise together to make something new? On a related tangent, I started thinking about depth; about how the relationship between panels changes when you start to position them in three dimensions instead of two. What does it mean to be able to stand in the space between two panels? And how might that experience be transferred back from the physical world to that of the screen? I find myself today with more questions than answers. What a wonderful condition in which to try and write a conclusion.

* * *

It's right now and I'm writing the foreword to this book.

I'm sitting down at my computer and I'm thinking about what I've just written. About the last twelve years of my work in comics and how I've used space and made spaces. And what I'm thinking is this – I'm not ready to make any conclusions. There's still too much out there to explore. Too many questions left to ask and too many comics still needing to be made. I open a notebook to an empty page and start to doodle a new configuration of panels across the page. I stare at the arrangement of little blank boxes in front of me. Perhaps they could contain anything, these boxes. But I can see a shape to them. I can see a pattern and a story and a potential laid out before me. I know exactly what they need to contain.

I pick up the pencil and I start to draw.

Jason Dittmer

INTRODUCTION TO COMIC BOOK GEOGRAPHIES

THE DIVIDES OF INTERDISCIPLINARITY

This volume is an attempt to bring together two fields that have never properly been in dialogue before: the discipline of geography and the interdisciplinary and nascent field of comics studies. While comics studies is indeed interdisciplinary, it tends to fall into several disciplinary traditions that all too rarely intersect with one another (e.g., the literary, the visual/aesthetic, cultural studies). This has both led to replication of work and a lot of talking past one another. This book is an explicit attempt to find common ground among scholars of comics studies through conceptual reflection: "comics studies, to thrive, must find a stable conceptual basis that is in no way interchangeable with conventional disciplinarity," (HATFIELD 2005, 13).

The division between geography and wider comics studies is all the more surprising given their mutual interest in space. For geography, spatiality has been a central concept for as long as there has been a discipline. Past visions of space as a container for social organization and action (HARTSHORNE 1939) have given way to more nuanced ideas of space as a phenomenon far from absolute in nature. Instead, recent theorists have turned to the ways in which space-time is produced through human relations in complex ways, drawing particularly from poststructuralist theorists such as Michel Foucault, Gilles Deleuze, and Jacques Lacan. This shift has coincided with the emergence of a 'spatial turn' throughout the humanities and social sciences over the past two decades, which has led to a fundamental re-imagining of the ways in which society operates. As Nigel THRIFT (2006, 139) writes: "The world is made up of all kinds of things brought in to relation with one another by this universe of spaces through a continuous and largely involuntary process of encounter[.]"

If ever there was a visual form well suited to speak to a universe of spaces which are brought into relation with one another, it is that of comic books. Scholars of the comics medium have extensively explored its formal properties, with early innovation taking the form of semiotic approaches in the Francophone world, several decades prior to Scott MCCLOUD's transfer of the topic to the Anglophone world via his tremendously influential *Understanding Comics* (1993, 67). "Comics panels fracture both time and space, offering a jagged, staccato rhythm of unconnected moments." While the formal analysis of comics has surpassed McCloud's initial attempt in the intervening two decades, especially with the translation of GROENSTEEN (2009) and his peers into English, MCCLOUD's general formulation of comics as the mapping of time remains influential.

In this introduction to the volume I would like to briefly trace the ways in which geography and comics are intertwined – including and beyond the semiotic – before returning to the foreword to see how space, place, and comics are entwined in Daniel Merlin Goodbrey's personal narrative. The chapter will conclude by summarizing each chapter and articulating my hopes for future interdisciplinary collaboration around spatial theory. I hope that you enjoy it.

SPACE, PLACE, AND COMICS

Helpful to me in formulating the relationship of the comics medium to geography has been the work of Paul ADAMS (2011), whose taxonomy of media geography fruitfully brings out the interrelations between space, place, and medium. ADAMS imagines these interrelationships in four ways: place in media, space in media, media in space, and media in place. In doing so, he acknowledges the relational turn in spatial theory described above by positing a world of motion, fluidity, and exchange (space) alongside a world of human experience, seeming stability, and rootedness (place). Of course, these are not mutually exclusive ontologies, but simply two different ways to grasp the process of becoming. The distinction between space and place is heuristic, an attempt to isolate something specific about the role of geography in constituting social relations.

The second dimension to ADAMS's typology is another tension – between the *context* and *contents* of any mediation – which intersects with space and place in complex ways.

> Space is where communication happens but is also one of the things created by communications, and, as CRESSWELL (2004: 15) writes, 'place is not simply something to be observed, researched and written about but is itself part of the way we see, research and write'. This paradoxical condition of mutual containment resists the imposition of hierarchical models[.] (ADAMS 2011, 39)

The role of media as something that both composes space and takes place requires a range of academic approaches to try and come to grips with the totality of a single medium. ADAMS categorizes these approaches according to their stances on these two tensions: space/place and context/contents. However, he is clear that no media event occurs only in one 'quadrant' of his typology. Here I will consider his typology with regard to the specific medium of comics.

'Place in comics' refers to the role of comics' representations in the constitution of particular places. It has long been understood that places are not isolated, autochtonous localities defined by their cultural characteristics. Rather, places are themselves produced through relations – not only among the people who live there but also the media objects that circulate in *other* places but nevertheless help to construct that place's identity. One need only think of tourism advertising, which is never found *in* the place that is being advertised, but which not only produces a place 'brand' but also shapes the experience of those who go there on vacation via the management of expectations.

With regard to comics, many sorts of places are of course represented from all over the world, but this is an uneven geography, not only in terms of *which* places are represented but also *how* they are represented. This is indubitably tied to the major hearths of comics production around the world. While comics are produced in many different places, the three most significant hearths are the United States, France/Belgium, and Japan/Korea. This fact skews the representation of place towards the cultural biases of these hearths. While it is a mistake to take American superhero comics as indicative of the entire medium, it is worthwhile to peruse some of the meta-texts produced by Marvel Comics to see an example of how geographic imaginations are embedded into comics. The *Marvel Atlas* (HOSKIN et al. 2008) maps the intersection between the 'our' world and the Earth of the Marvel Universe. The *Atlas* has entries for many countries in our world, tracing what battles were in which countries, and which heroes and villains originated where. More interesting, perhaps, are the imaginary countries which have been invented by writers over the years; these countries – like Wakanda and Madripoor – might be understood as regional stereotypes. Intriguingly, they cluster in a few regions which serve as exotic frontiers for superheroic intervention: east and southern Africa, the Balkans, Central Asia. Such representations – of specific countries and of broader regions – help to compose places in the 'real' world. This reliance on stereotype and thin description can be contrasted with the *Marvel Comics Guide to New York City* (SANDERSON 2007) which likewise conducts a survey – often block-by-block and building by building – of the intersection between the Marvel Universe and the urban stage for most of its action. The depth of lived experience among most of Marvel's creative staff in the Big Apple means that Marvel's comics detail a rich urban geography which is only augmented by its fantastic elements. Similar processes of cultural myopia can be found in Japanese manga or Francophone *bandes dessinées* (as we shall see in Catriona MacLeod's chapter).

'Space in comics' refers to the role of space in the *content* of comics. Although ADAMS refers to this as the least fleshed out quadrant of his typology in the media geography literature, it is – thankfully – one of the most considered aspects of comics studies. A substantial literature has emerged, from the already mentioned MCCLOUD and GROENSTEEN to more recent interventions by Charles HATFIELD (2005) and Douglas WOLK (2008), examining the formal properties of comics and how they produce narrative. Whether in the Francophone visual tradition (GROENSTEEN's concept of spatio-topia) or the more literary conceptualizations of comics (HATFIELD's series of tensions), each shares an interest in the way that images and text in comics are alchemically set in relation to one another and therefore compose a topological space (see DITTMER 2010; GALLACHER 2011 – but for a contrarian view, see Marcus Doel's chapter in this volume).

Importantly, this topological understanding of comics not only helps us to understand the way a single panel, or a single page, works, it also helps us to view the larger set of connections that run through the archive. This is more than just the occasional visual homage or inside joke (although it is those too). Serial narratives, which most comics are (but certainly not all), rely on notions of continuity and intertextuality that are not unique to comics but which are certainly very intensely layered. Meta-texts like the Marvel guide to New York City and indeed the world, described above, rely on these topological webs of relations that collectively enable the production of specific places. Comparing space in comics and place in comics enables us to grasp, perhaps for the first time, the way in which each grasps at some dimension of the role of geography in social life. Each contains the other within it.

Comics in space refers to the various ways in which comics can be circulated, and how this opens up new possibilities for the form. Some have argued for a very tight definition of comics, rooted in their industrial emergence and production. This view emphasizes the materiality of comics as a particular kind of object (paper in a booklet form) in a particular mode of distribution (physical transportation by mail or through commercial distribution networks). The advent of web-comics was exciting precisely because it allowed comics to be reimagined as something beyond this materiality (McCloud 2000). Intriguingly, site-specific comics have emerged as the antithesis of the dematerialized web-comic, requiring the 'reader' to physically go to the site and experience the comic there (Dittmer 2011). Pushing this concept further, it becomes possible to imagine, for example, a city as a comic (Ahrens and Meteling 2010) or to 'read' geographies in the relationality of space (as we will see in Tony Venezia's chapter in this book). Collectively, the notion of comics in space highlights how comics circulate (or do not) and how that 'how' shapes the comic itself.

The final quadrant of the typology is comics in place, and it refers to the way in which the location of comics – the context of reading, or production, matters. This is a long-standing interest in comics studies, with Martin Barker's landmark *Comics: Ideology, Power, and the Critics* (1989) pointing away from textual analysis to a concern with the context of consumption. This dovetails with a lot of work in geography that looks at the circulation of texts and their re-interpretation in different contexts (Livingstone 2005). Similarly, economic geographers have looked at changes in the production of comics (Norcliffe and Rendace 2003), arguing that the decentralization of the comics industry (assisted by the internet) has opened up the range of places from which writers and artists can create. The impact of place on comics production is considered in Shaun Huston's documentary *Comic Book City, Portland, USA* – a topic which Huston also illuminates for us later in this volume.

Together, these four interrelations between space/place and context/contents point to a conceptual language that provides a firm grounding for a joint exploration of comics by scholars in geography and the wider field of comics studies. However, before we get to the authors I would like to revisit Daniel Merlin Goodbrey's foreword in order to show how this typology helps us to make sense of his own personal narrative of a life in comics.

BACKWARD TO THE FOREWORD

I first saw Daniel's work at the *Hypercomics* exhibit in which he contributed his work *The Archivist* (described in the foreword). I was captivated by his work and that of the other artists on display. It was clear that Daniel was thinking about comics, space, and place and their multiple intertwinings in precisely the ways that I too was exploring. While his work is practice-led (although engaged with academic literature) and I have no artistic skills whatsoever (sigh), we shared a common interest in the contingency of reading practices, the embodiment of the viewer/reader, and the relational space both between panels and between the panels and 'real' space. When I asked Daniel to offer the foreword to this book, I charged him simply to discuss the relationship between space and comics, and I am thrilled with what he sent in.

Daniel's narrative pulls together many of ADAMS's ideas regarding the space/place and contents/context tensions that I have just outlined, in a much more satisfying form than I was able to produce. His main argument is that his artistic career has been a progressive attempt to rework the relationships between comics in space and space in comics. His shift first from the page to the computer, and then from the computer to the gallery space (never entirely leaving past forms behind), is not just an effort to move comics around (or fix them in place) but also, simultaneously, an effort to push the limits of the aesthetics and topological elements of comics themselves. He notes first how the shape and materiality of the page pose certain limits on the artist, but that those limits are also potentially creative in nature; the layout of the page can be sculpted to shape a narrative and its pacing, which in turn can have effects in the non-narrative world – an exhausted artist filling lots of small boxes, or a reader who is affectively drafted along in the wake of an end-of-page cliff hanger. The shift to digital comics frees the artist of these limitations, making space potentially infinitely stretchable. This plasticity however looses the moors of convention and potentially disorients both artists and readers who grew up with certain habits of narrative. I recall a 1994 episode of the Simpsons ("Deep Space Homer") in which Homer accidentally smashes an ant farm, and as the ants escape they exclaim "Freedom! The horrible, horrible freedom!" Luckily for comics artists, the computer screen itself provided a material constraint (like Homer's ant colony, one made of framed glass) on the use of space in comics, requiring artists to think through the material constraints on how their work would be viewed, even if they were working on a so-called 'infinite canvas'.

The next shift in Daniel's career saw him taking up the space of the gallery as a laboratory for examining space in comics, but here the tension with comics in place comes out strongly as well. For instance, his observation that the spatial orientation of the narrative would have to map onto the direction of foot traffic indicates the importance of embodied habits that are rooted in a particular site and its architecture (it is notable that his *Casita Situations* exhibit was collaborative with an architect). Further, the place-based nature of site-specific works such as *Casita Situations* and *The Archivist* connect not only with the embodied viewer but also with the richness of their locales: *Casita Situations* was designed for a young people's mental health clinic (to produce a therapeutic atmosphere) and *The Archivist* was meant to engage with the Pump House Gallery and Battersea Park more broadly, whether through the landscape or through (as Daniel did) a fictional history of the site. Of course, that means that *The Archivist* also represented the Pump House Gallery within its narrative, linking to the notion of place in comics. One might also consider his meta-textual inclusion of his comics' flow diagrams in the comics themselves (recall his comic *Six-Gun*) as a hitherto unpredicted inclusion of 'space in comics in comics'! By now it should be clear why I find Daniel's work so compelling and why I am so grateful to have him in this volume.

THE CHAPTERS

Speaking of this volume, it is now time to turn to the other contributors. Their work all draws on one or more aspects of ADAMS's typology. This book is divided into three parts, although the boundaries between these parts are quite permeable and some chapters could easily be located elsewhere. The first section is 'Representing and performing place/space', and it includes chapters by Oliver Dunnett, Tony Venezia, and Shaun Huston. These chapters all centre on the role of comics in the everyday production of places.

Oliver's chapter, 'Framing landscape: *Dan Dare*, the *Eagle* and post-war culture in Britain', draws on the landscape tradition in cultural geography and uses it to illuminate a historical debate over the role of the British countryside in the formation of national identity in the post-war era. Oliver does this by engaging in a visual analysis of Dan Dare comics in the *Eagle*, a key British comic in the mid twentieth century. The tales of space pilot Dan Dare, Oliver argues, incorporated nostalgic visions of the British countryside into a modern, progressive narrative of continued British geopolitical relevance. While his visual analysis is excellent, I particularly like Oliver's detailed historical contextualization of the foundation of the *Eagle* and how it shaped the narratives that it told.

Tony's chapter, '10th April, 1999, Conway Hall, Red Lion Square: *Snakes and Ladders*, occult cartography and radical nostalgia', is awkwardly titled but richly rewarding. In it, Tony looks to Alan Moore's magical performances, which draw on the tradition of psychogeography to highlight the occult relations between sites in London. While this might be understood as a form of comic-at-its-limits (see my earlier comments about comics in space), Tony's task is aided by Eddie Campbell's comic, *Snakes and Ladders*, which adapts the live performance of magical psychogeography into graphic narrative. This allows Tony to reflect on the formal properties of comics and how they intersect with psychogeography, nostalgia, and relational spaces. Particularly pleasurable in this chapter is Tony's writing style, which is a bit occulted itself – meriting multiple readings to pull out the rich meanings inside.

'Live/Work: Portland, Oregon as a place for comics production', Shaun's chapter, is markedly different than all the others in this volume. It focuses not on comics at all, but on the way that comics culture has found a particular home in Portland. Portland is home to several production companies and a rich independent comics scene. Shaun reports back from interviews with these creators to understand their motivations for moving to Portland and staying there. Connecting these findings to broader literature in urban and economic geography (notably the 'creative class' literature), Shaun finds that 'creativity in place' is likely to remain a key feature of Portland for years to come. Shaun's chapter not only speaks to a side of comics that is rarely addressed, but he has also produced a film (*Comic Book City*) about the topic. Intriguingly, the film attempts to incorporate the space of comics into the visual style of the comics, exhibiting meta-textual qualities like Daniel Merlin Goodbrey's *Six-Gun*.[1]

The second part of this volume is entitled 'Bodies politic'. It features chapters by Catriona MacLeod, Juliet Fall, and Ted Holland. These chapters are united by their concern for politics and the role of bodies in constituting subjectivities, but they do so in very different ways despite two of the chapters being notionally very similar. Catriona's chapter, 'From wandering women to fixed females: Relations of gendered movement through post-colonial space in *Lettres d'Outreimer* and *Le Bar du vieux Français*' compares two different *bandes dessinées* written about postcolonial contexts – one in Morocco and one in the French Caribbean. Catriona's central interest in the role of women's bodies in the scripting of both place and space – place in terms of the representations of these contexts, and space via attention to the (im)mobility of these women's bodies. The chapter is connected with, and contributes to, the literature on migration and postcolonialism. Catriona's chapter is buoyed along by her close, and yet still highly narrative, reading of the comics; she manages to hold together the postcolonial, gendered, and spatial elements of the comics in a highly readable way.

Juliet's chapter is entitled 'Put your body on the line: autobiographical comics and the shared ironies of bounded lives' and it is a sophisticated take on the recent boom in autobiographical comics, such as *Persepolis*. In particular she looks to three authors – Joe Sacco, Guy Delisle, and Maximilian Le Roy – and draws on the work of Judith Butler in order to see how autobiographical comics can contribute to the forging of connections with distant others. A common thread through these works is a sense of embodied precarity – not for the creators themselves but for those they encounter – and the various ways in which the creators cultivate this sense through plurivocal narration is the centrepiece of Fall's analysis. What I enjoy about Juliet's chapter is the way that she weaves the comics' narratives into her argument, thereby cultivating the very same empathy that she is studying in other's works.

Ted's chapter also deals with the work of Joe Sacco. In 'Post-modern witness: Journalism and representation in Joe Sacco's "Christmas with Karadžić"', Ted

1 The film website is http://olympus_mons.typepad.com/pdxcomicsdoc/

drills down into a single short story, first showing its connections to the wider Bosnian conflict on which Sacco was reporting, and then highlighting the way that Sacco's positionality as a comics *journalist* shaped his encounter with notorious military commander Radovan Karadžić. Ted, like Sacco himself, is interested in different narratives that might work against violence and oppression, but he worries that not all 'grounded' narratives can work against these forces because encounters with the other are always structured by pre-existing subjectivities. Ted's chapter is an important reminder that while comics offer different opportunities than other media, there is no reason to believe that they can escape many of the same limits that constrain other narratives.

The third section of the book is entitled 'Space and comics theory', and it includes chapters by Julia Round, Mitch Goodrum, and Marcus Doel. Several of these chapters could be found elsewhere, and indeed others could fall into this section. However, these were selected as they most explicitly consider the spaces in comics – the way in which comics work to produce narrative, and how that can be pushed to tell particular kinds of stories. I concluded the volume with this section because the diversity of approaches to space in comics (literary theory, psychoanalytic theory, and poststructuralist philosophy) indicates how fruitful and varied an on-going dialogue between comics and spatial theory might be.

Julia's chapter – 'We share our mother's health: temporality and the gothic in comic book landscapes' – links the visual landscapes of *From Hell*, *The Invisibles*, and *The Walking Dead* with the much-studied relationship between time and space in comics. It does this, intriguingly, via gothic theory. Using concepts such as abjection, fragmentation, inversion, and cryptomimesis, the creators of these comics tie together a range of temporalities in single sites that are all-at-once. There are obvious overlaps with Tony's chapter on psychogeography, not only through Alan Moore (the author of *From Hell*) but also through both author's emphasis on the occulted city. I enjoy Julia's chapter for many reasons, but particularly for her well-considered applications of GROENSTEEN's comics theory to the gothic.

Mitch's chapter also uses the concept of fragmentation, introduced by Julia, but takes it in a different direction. In 'The body (politic) in pieces: Post 9/11 Marvel superhero narratives and fragmentation', Mitch considers several crossover narratives produced by Marvel Comics that feature virtually all of their characters: *Avengers Disassembled*, *House of M*, *Civil War*, and *Secret Invasion*. Mitch reads these as commentaries on the superhero genre in the wake of 9/11, but his analysis pushes further than this simple geopolitical link. Rather, he reads the visual tropes of these story arcs through the lens of psychoanalysis; the trauma of 9/11 leads to the fragmentation of superhero narratives, something visible not only in the narrative dissolution of superhero teams and even superhero realities but also in the fragmentation of images. These images pick up on the tension between a comic as a collection of panels and a comic as singular narrative and follow the same processes of deterritorialization and reterritorialization. I really enjoy Mitch's chapter because it cleverly connects these formal elements of comics and connects it with the pro-social mission of superheroes; both involve fragmentation and eventual reconstitution.

The final chapter of the volume belongs to Marcus Doel. Marcus's title is too long to introduce here given stringent word limits, but is available for perusal in the table of contents. Marcus's chapter is a contribution to comics theory, supplemented by references to Chris Ware's oeuvre. Drawing on poststructuralist philosophy about the nature of time and space, Doel offers a broadside against conventional comics theory in which he contests the nature of comics as sequential, braided, or any other form of topology. His call for 'some comics' rather than any sense of finitude will prove provocative to many in the field, I suspect. I found Marcus's chapter challenging but well worth the effort to read and re-read; his weaving together of Ware's old-fashioned transitions ("AND SO") with his larger argument lends an air of exuberance to a dense yet meaningful argument.

With that, I am ready to hand the baton on to these authors, who have so generously contributed their time and intellectual energy to this book. It is my hope that this book should prompt a greater engagement between geographers and comics studies scholars around spatial theory. In this introduction I have tried to lay the groundwork for such a project, and I have purposely left the disciplinary identities of each author to the side; all the better to jump start a truly interdisciplinary engagement.

REFERENCES

ADAMS, P. (2011): A taxonomy for communication geography. *Progress in Human Geography* 35, 37–57.

AHRENS, J. and A. METELING (Eds.) (2010): *Comics and the City: Urban Space in Print, Picture, and Sequence*. London and New York.

BARKER, M. (1989): *Comics: Ideology, Power, and the Critics*. Manchester.

CRESSWELL, T. (2004): *Place: A Short Introduction*. Malden, MA.

DITTMER, J. (2010): Comic book visualities: a methodological manifesto on geography, montage and narration. *Transactions of the Institute of British Geographers* 35, 222–236.

DITTMER, J. (2011): "*The Rut* by Dave McKean" (Art exhibit review). *Studies in Comics* 2, 380–387.

GALLACHER, L.-A. (2011): (Fullmetal) alchemy: the monstrosity of reading words and pictures in shonen manga. *Cultural Geographies* 18, 457–473.

GROENSTEEN, T. (2009): *The System of Comics*. Jackson, MS.

HARTSHORNE, R. (1939): *The Nature of Geography*. Lancaster, PA.

HATFIELD, C. (2005): *Alternative Comics: An emerging literature*. Jackson, MS.

HOSKIN, M., S. VANDAL, A. FLAMINI and E. MOREELS. (2008): *Marvel Atlas*. New York.

LIVINGSTONE, D. (2005): Science, text and space: thoughts on the geography of reading. *Transactions of the Institute of British Geographers* 30, 391–401.

MCCLOUD, S. (1993): *Understanding Comics: The invisible art*. New York.

MCCLOUD, S. (2000): *Reinventing Comics*. New York.

NORCLIFFE, G. and O. RENDACE (2003): New Geographies of Comic Book Production in North America: The New Artisan, Distancing, and the Periodic Social Economy. *Economic Geography* 79, 241–263.
SANDERSON, P. (2007): *Marvel Comics Guide to New York City*. New York.
THRIFT, N. (2006): Space. *Theory, Culture, and Society* 23, 139–146.
WOLK, D. (2008): *Reading Comics: How Graphic Novels Work and What They Mean*. Cambridge.

PART 1 | REPRESENTING AND PERFORMING PLACE/SPACE

Chapter 1 | *Oliver Dunnett*

FRAMING LANDSCAPE: *DAN DARE*, THE *EAGLE* AND POST-WAR CULTURE IN BRITAIN

The *Eagle* was one of the most popular post-war children's papers in Britain, and at its peak in the 1950s it attracted almost one million readers each week (MORRIS 1977). Recently described as "arguably the most historically significant British comic" (CHAPMAN 2011, 45), *Eagle*, along with its star attraction Dan Dare, still looms large in the imaginations of its former readers and holds a prominent status in British cultural history. Created by Reverend Marcus MORRIS (1915–1989), who worked closely with principal illustrator Frank HAMPSON (1918–1985), the paper was first published on 14 April 1950, with regular features including a range of comic strips, editorials, factual articles, spy stories and sports pages.[1] *Eagle* sought to set itself against both the established canon of children's papers as well as the growing number of publications being imported to Britain, through the presentation of a strong moral and ethical tone set alongside a vision of technological optimism, themes that were conveyed most clearly through HAMPSON's space adventure comics, *Dan Dare – Pilot of the Future*. Through focusing on the formative backgrounds and representational strategies of *Dan Dare and the Eagle*, this chapter examines the ways in which HAMPSON and MORRIS encapsulated and sought to shape a certain vision of British society in the early post-war period.

COMICS, GEOGRAPHY AND NATIONAL IDENTITY IN POST-WAR BRITAIN

In examining MORRIS and HAMPSON's vision of post-war Britain, this chapter considers some of the ways in which geographies of the nation-state have been manifested in comics. Whilst comics studies and geography have not to date overlapped by a great amount, research in political geography has drawn attention to the wider significance of popular culture in defining aspects of national identity, including in lifestyle magazines such as *Reader's Digest* (SHARP 2000), and indeed in comics such as *Captain America* (DITTMER 2005) and *Tintin* (DUNNETT 2009). Such studies have shown how some of the analytical techniques of contemporary human geography can be applied to the critical interpretation of comics. This chapter adds to this productive vein of work by adopting a cultural land-

1 MORRIS edited *Eagle* until 1960, and this chapter focuses on extracts from the paper's first few years. *Eagle* was eventually absorbed by *Lion* in 1969.

scape perspective in its interpretation of *Dan Dare* and the *Eagle*. This approach has underpinned a great deal of work in critical human geography, understanding landscape as "a cultural image, a pictorial way of representing, structuring or symbolising surroundings" (DANIELS and COSGROVE 1988, 1). By examining cultural landscapes in this way, insights from cultural geography can be added to the toolbox of comics scholarship alongside existing approaches pioneered by EISNER (1985), MCCLOUD (1993) and others. At the same time, this approach brings comics and their unique representational strategies further into the fold of cultural, historical and political geography, which have until recently only engaged with comics on a limited scale.

A particular historical issue upon which this chapter may also shed light is a debate that surrounds the characterisation of British national identity in the post-war period. Here, some have pressed the view that Britain at this time was entering a period of terminal decline fuelled by an anti-capitalist shift in society (WEINER 1981; BARNETT 1986), whilst others have argued that developments in areas such as the aeroplane industry represented "the modern side of England's heritage" (EDGERTON 1991, 90), with each point of view invested in certain ideas of landscape and national identity. Central to this debate have been the ways in which 'declinist' writers have criticised traditional representations of British heritage, which they say were still a dominant motif in post-war society, and characteristic of an anti-technological society. Indeed, HEWISON (1997) has noted that a nostalgic view of the national heartland, represented by the English countryside, had become a central expression of national identity in the 1940s, and that this helped stifle the nation's sense of progress in the post-war period. By contrast, WEIGHT (2002, 159) has suggested that this conception of English rural Romanticism has been "grossly exaggerated", whilst MATLESS (1998) has identified a more progressive view of landscape in the mid-century period, which held the modern and the traditional in alignment, as complementary elements of national identity. One example of an under-represented English rural modernity was electricity supplied by the National Grid, whose steel pylons were erected across parts of the countryside in the late-1920s and 1930s, and were seen by some as "part of a preserved, modern, rural England [which] carried aesthetic as well as social Power Potential" (MATLESS 1998, 52).

As well as focusing on representations of the English countryside, this debate also crystallised through contrasting interpretations of the 1951 Festival of Britain, whose centrepiece exhibition at London's South Bank was intended to paint a picture of British culture in the past, present and future, and became well-known for its futuristic-looking 'Dome of Discovery' and the still-surviving Royal Festival Hall (CONEKIN, 2003). Hence, whilst some writers have noted that "against the background of thriving tradition, the Festival celebrated modern industry and technology" (WEIGHT 2002, 198–199), others maintain that "the modernist architecture was a lightweight framework for yet another exploration of Deep England" (HEWISON 1997, 59). In a similar way to the Festival, the *Eagle* has been adopted by some writers as an indicator of post-war British culture (EDWARD 1987; SILLARS 1995). Indeed, one critic has suggested that, whilst the *Eagle* was "formed within the historical conditions and ideological discourses of post-war Britain", it can be seen as "a vehicle for the promotion of consensus", indicative of conservative values and a middle-class ethos (CHAPMAN 2011, 53, 58). As such, the *Eagle* offers interesting perspectives on the issue of national decline and the associated discourses of technology, landscape and culture. Whilst these themes shall be returned to in the conclusion, what follows is divided into two sections; the first providing an overview of the *Eagle* and the background to the creation of *Dan Dare*, and the second focusing more closely on the representation of landscape in these comics.

THE DEVELOPMENT OF *EAGLE* AND *DAN DARE*

Marcus MORRIS began his career as a Lancashire vicar in the 1940s, and his interest in parish magazines first led him into the field of publishing. Having felt that these publications were often "dreary and ineffective", he set about converting a four-page parish leaflet into a magazine called *The Anvil*, in which issues of the day would be "hammered out" in an attempt to convey "an intelligent view of Christianity" (MORRIS 1977, 3). He then became interested in developing a different kind of publication, which would be aimed at children, but was still envisaged through the Christian moral lens which was so important to MORRIS. Indeed, it has been noted that the name of the new paper was inspired by an eagle-shaped Bible lectern (CROMPTON 1985), in what was an appropriate metaphor for the *Eagle*'s tone of delivery; dynamic and exciting, but grounded in an ethos of Christian morality.

Moreover, a key motivation for MORRIS was to counteract what was seen as a proliferation of low-brow imported American comics. These were at the time arriving from across the Atlantic as ships' ballast, and they helped comics develop a reputation in Britain as "bad mass-art geared to a very low mental age" (HOGGART 1957, 153). Indeed, MORRIS wrote an article for the newspaper *Sunday Dispatch* railing against such publications and their perceived negative effects on children (MORRIS 1949), and as such, he contributed to the 'horror comics' backlash that eventually led to the banning of certain publications in Britain under the Children and Young Persons (Harmful Publications) Act of 1955. The problem, as perceived by certain sections of British society, was "American imperialism … by which every possible bad facet of American life was being exported", whilst the solution was "the promotion of better books and magazines … national decency and higher values" (BARKER 1984, 23–26). Indeed, MORRIS was convinced that "the strip cartoon was capable of development in a way not yet seen in England", and "could be used to convey to the child the right kind of standards, values and attitudes, combined with the necessary amount of excitement and adventure" (MORRIS 1977, 3).

MORRIS had recruited Frank HAMPSON from Southport School of Arts and Crafts as an illustrator for *The Anvil* in 1948, and having decided on their ambitious new venture, the drive to publish the first copy of *Eagle* began in 1949. After being turned away from established comics publisher Amalgamated Press and other leading publishers in London's Fleet Street, MORRIS eventually found favour with Hulton Press (MORRIS 1977). This relatively small firm, which also produced the left-leaning photojournalistic magazine *Picture Post* and the art and literature magazine *Lilliput,* was perhaps in a better position to accept *Eagle* on its own terms, instead of in comparison to the established British comics of the time. As such, MORRIS and HAMPSON were able to develop *Eagle* not only as a reaction to the 'horror comics', but also as a quite radical departure from the existing British tradition of comics. These were typified by titles for younger children such as *Dandy* and *Beano*, against which *Eagle* would be presented as more realistic and intellectually challenging. Indeed, it has been noted that the publishing industry in the early 1950s was no less than "overhauled" by MORRIS and HAMPSON's efforts (GRAVETT and STANBURY 2006, 13). As this implies, *Eagle* was an instant success, the first issue being a sell-out of 900,000 copies. In an unusual strategy for a children's paper, MORRIS outlined a kind of manifesto in his first editorial, where he described *Eagle* as "an entirely new kind of strip-cartoon paper." As part of a concerted enrolment strategy, MORRIS invited readers to join the Eagle Club, which was to be "one of the most important features of the paper [with] very definite aims and standards" (MORRIS 1950, 3). The Eagle Club drew in applications from around 60,000 readers after the first two issues, and members were encouraged to behave in certain ways, including an expectation to "develop themselves in body, mind and spirit" (MORRIS 1950, 3).

Alongside messages about society and good character promoted directly by MORRIS in his editorials, and which were implicit in some of the paper's other features, a further major thematic element of *Eagle* was the representation of modern technology. Exemplifying this theme, the large 'cutaway' pictures by artist Leslie ASHWELL WOOD held a prominent status at the centre fold of each weekly issue. These pages presented a striking double-page image, usually either an example of new civic amenities, such as the Festival of Britain's Dome of Discovery, or the latest modern technology, such as the artificial satellites of the late 1950s. ASHWELL WOOD'S illustrations have been described as "not only artistically clean and sharp but also educational, offering clear explanations to complicated scenes" (National Archives 2012, n.p.). Their distinguishing feature was that some of the outer layers of the object in question would be cut away, revealing its inner workings, whilst embedded reference numbers were linked to detailed footnotes. In this way, the reader is assigned the privileged perspective of a scientist or engineer, and is presented with pictorial as well as written explanations to a level of detail that would otherwise be unseen. ASHWELL WOOD'S illustrations also appeared in wartime leaflets and pre-war popular science magazines such as *Modern Wonder,* and can be considered part of an established tradition of illustrations and technical diagrams which contributed to a significant popular science movement in pre-war Britain (BOWLER 2006). In exemplifying *Eagle*'s theme of optimistic technological realism, this kind of feature sits alongside the paper's presentation of comics, adding to the sense that readers would be technically-minded and keen to embrace the realistic, detailed environments presented in comics such as *Dan Dare*.

Frank HAMPSON's *Dan Dare* comic strip was the flagship feature of *Eagle*, occupying the first two pages of each issue. These space adventure comics, set in a future vision of the 1990s, were known for their vibrancy, use of colour and attention to detail, which was in line with MORRIS's desire to produce a high quality publication for children. This standard was achieved through the use of a studio system, whereby several artists worked on the strips simultaneously, and items such as military uniforms and space ships were often copied either from models or the real thing. A further level of scientific rigour was added with the employment of science writer Arthur C. Clarke as an adviser for some of the early editions of *Dan Dare*, which involved him contributing to the scripts as well as checking the drafts for technical accuracy (MCALEER 2002). Clarke, a space flight visionary and post-war Chairman of the British Interplanetary Society, had written to MORRIS in the spring of 1950:

> I think this might amuse you. Yesterday I was lecturing at the Royal Geographical Society on the problem of interplanetary navigation ... After a highly technical series of remarks, [one of the other speakers] ended up by asking "Will Dan Dare reach Venus?" (quoted in MORRIS and HALLWOOD 1998, 141).

This mention of *Dan Dare* refers to a narrative climax from one of the opening episodes, which was publicised at the time on billboard posters, and gives a good sense of the broader public awareness of *Dan Dare* over and above its target audience of children. Following his close involvement with MORRIS and HAMPSON during 1950, Clarke is said to have thought that "the standard of work and research [in *Dan Dare*] was so high that they were wasting their money getting him to check it" (VINCE 2007, 37), and this particular collaboration came to a close, although Clarke later contributed a short science fiction story to *Eagle* under a pseudonym (WILLIS 1950). Indeed, the similarities between Clarke and HAMPSON's science fiction have been picked up on by commentators such as James EDWARD (1987), who identifies optimism in their work that was at odds with most of the other science fiction being produced in Britain at this time.

In order to fully appreciate the representational strategies in *Dan Dare*, it is necessary to gain an understanding of the identity of the eponymous space pilot. According to MCCLOUD's (1993) typology, the pictorial rendering of cartoon characters in comics varies on a scale from the realistic, representing a character in all their detail, to the iconic, where it is the reader who imaginatively fills in the gaps, and ultimately assumes the identity of the character. An initial look at the form of Dan Dare suggests that he falls more towards the realistic end of this scale, with his famous 'lantern jaw', jagged eyebrow, green military uniform and pilot's cap. However, identity in comics is not just bound up with graphical depiction, and through closer analysis we can identify a whole range of ideas which influenced the quite specific character of *Eagle*'s main protagonist, ranging from the pervading ethos of Christian morality with which the paper as a whole sought to define itself, through to the direct personal experiences of HAMPSON himself.

HAMPSON and MORRIS's initial proposal for their star attraction was a character called Lex Christian, a "tough, fighting parson in the slums of the East End of London" (MORRIS 1977, 4). This type of moral heroic identity is closely aligned with what MORRIS saw as a necessary antidote to the imported horror comics, and also bears the reader back to the traditional adventure hero of the nineteenth century boys' adventure story (WATKINS 2000). Furthermore, it appears that the fantasy writer C S Lewis's little-known space trilogy was a significant influence on both the character of Dan Dare and his primary destination in outer space. Although more widely known for his *Narnia* series of fantasy novels, these earlier works of Lewis, first published between 1938 and 1945, have since been brought into the fold of academic critique and have been considered as part of a wider tradition of Romantic protest in Britain (VELDMAN 1994). MORRIS was an admirer of Lewis, who was well-known at the time as a Christian apologist, and indeed Lewis once wrote an article for MORRIS's aforementioned parish magazine (CROMPTON 1985). MORRIS was particularly struck by the second instalment of Lewis's trilogy, *Perelandra*, which is an allegory of the Garden of Eden story set on the planet Venus. MORRIS explained that "one day, after re-reading C S Lewis's novel *Perelandra*, I said to Frank [HAMPSON] that I thought Lex Christian should leave London and go out into space" (MORRIS and HALLWOOD 1998, 100).

Whilst MORRIS would have been inspired by this type of popular adventure story, with the Christian values that he found so lacking in popular culture, he saw space exploration in a positive light, as opposed to Lewis, who considered it an ominous prospect (DOWNING 1992). Moreover, MORRIS and HAMPSON understood the excitement of space exploration, and foresaw the enthusiasm with which children would take up the idea. Having decided on the space-bound destiny of their new character, Lex Christian soon became "a flying padre, the Parson of the Fighting Seventh" (MORRIS 1977, 4), before turning into an interplanetary 'Chaplain Dan Dare', complete with collar, white cloak and gloves (MORRIS and HALLWOOD 1998).

However, in later drafts it was decided to drop any explicit references to Christianity in the character of Dan Dare, perhaps in light of the decision to run several Bible-related stories in other parts of *Eagle,* or as part of a strategy to reach out to as broad an audience as possible through the paper's front-page feature. Also related to the wider moral message of *Eagle* is that Dan is not a type of 'superhero' character, and does not possess any special powers; he simply uses his skills as a pilot to escape danger, and often seeks non-violent solutions to problems. Indeed, having been refined through various incarnations, Dan's principal identity is that of an ace pilot. Although he is arguably presented as more of a working-class hero, this has obvious connections in the 1950s to the idea of the popularly revered "heroic, aristocratic pilot" (EDGERTON 1991, 61) and the Churchillian notion of 'the few' airmen who won the Battle of Britain in 1940. This type of identity is mixed up with HAMPSON's personal experience of the Second World War, and, having unsuccessfully applied to be in the Royal Air Force, he admitted that Dan Dare "was the man [he] always wanted to be" (CROMPTON, 1985, 9). HAMPSON did, however, serve as an army truck driver in the war, was evacuated from Dunkirk in 1940, and later took part in the D-Day landings at Normandy. These heroics are, indeed, re-enacted in some of the early episodes of *Dan Dare.* HAMPSON's interest in space flight can also be traced to his wartime encounter of German rockets. Having reached Belgium in 1944, "he saw the doodle-bugs on their way to London", and later recalled that, "on the quays of Antwerp you could watch the birth of space travel" (CROMPTON 1985, 19).

Here we can see that Dan Dare in many ways embodied *Eagle*'s vision of sober Christian values elided with a thrilling sense of adventure that was partly drawn from a shared cultural reverence of the wartime pilot hero. These were perceived by MORRIS and HAMPSON as essentially British characteristics, and would serve to re-enforce a positive sense of national identity at a time when British values were in many ways considered under threat from American cultural hegemony (WEIGHT 2002). In any case, the theme of *Dan Dare* as an embodiment of both traditional values and modern progress is a theme that can be extended through examining the broader representational strategies of the series. Here, whereas some writers have suggested an incompatibility between archetypal images of British landscapes and the idea of progress, this closer analysis will support the argument that the progressive and traditional can be seen as complimentary elements of a positive vision of British post-war society.

LANDSCAPE AND CULTURE IN *DAN DARE*

The type of world which HAMPSON chose to convey in his vision of Britain in the 1990s was most clearly articulated through depictions of landscape in the early episodes of *Dan Dare.* Whilst primacy was given to pictorial representations wherever possible to drive the story, a certain amount of plot exposition was also imparted through wording in speech bubbles and text boxes. In this way, the narrative context of the first *Dan Dare* story is outlined, with one of Dan's colleagues explaining that "the number of people in the world has doubled since 1950 – and food supplies are growing less" (HAMPSON 1950a, 2). Here, two issues of the 1950s are compounded into a neo-Malthusian problem, which drives the storyline in this first *Dan Dare* adventure. Fears of over-population had been given a renewed sense of credibility in this period by scientists such as the British evolutionary biologist Julian HUXLEY (MONTGOMERY 1965), and, by 1950, food rationing had been in place in Britain for a decade, in some instances being even more severe than it had been during the war (KYNASTON 2007). This domestic phenomenon is reflected in HAMPSON's future vision of Britain, so instead of rationing books, mass-produced "vitamin blocks" induce in Dan and his friends the kind of complaints that would have been familiar to readers at the time (HAMPSON 1950b, 2). HAMPSON frames interplanetary exploration as a solution to food shortages and over-population, and Dan is sent to the planet Venus to seek out new sources of food, on a mission that "would mean life itself to millions of people" (HAMPSON 1950a, 2).

MORRIS has explained how he and his team of artists "tried to tell the stories mainly through the dramatic sequence of the pictures" (MORRIS 1977, 6). The ways in which these sequences are constructed provide insights into MORRIS and HAMPSON's perception of a future Britain, whilst at the same time, by drawing on the conventions of contemporary cultural geography, certain panels can be interpreted as cultural landscapes in their own right. This is particularly relevant to HAMPSON's artwork, in which the influence of his art school education can be seen through the use of traditional watercolour painting, with detail added in black ink outlining. By examining certain frames in detail, whilst also considering their sequential context, this section focuses on three scenarios from the first *Dan Dare* adventure which draw on some of the aesthetic and technological discourses of 1950s Britain in their depiction of a futuristic space port, a traditional English village, and a fly-by survey of the countryside.

In the very first frame of *Dan Dare*, the reader is introduced to "The Headquarters of the Interplanet Space Fleet" as part of a sequence that establishes this setting and depicts the launch of the 'Kingfisher' space ship that we can see in the centre of the frame (Figure 1.1). Whilst the presence of various rockets and flying machines form a particularly striking aspect to this frame, made all the more vivid by the dramatic aerial placement of the reader's eye, the details of the surrounding landscape also draw attention. HAMPSON's studio team put considerable effort into maintaining the accuracy and continuity of Dan Dare's launch pad to outer space as it was depicted from various different angles, to the extent that a complete scale model was constructed, with photographs then used as visual templates (CROMPTON 1985). Throughout this sequence, HAMPSON's modernist aesthetic is revealed in his choice of buildings, which are sleek, white and interspersed with greenery. These have been interpreted as "recognisable from recent British originals like Tecton's Finsbury Health Centre" (SILLARS 1995, 137), which was designed in the 1930s by the group responsible for London Zoo's penguin enclosure and other well-known modernist designs. Moreover, when viewing the space port landscape as a whole, as well as from different angles in other frames, the general impression is of an idealised modern industrial landscape that bears certain similarities to the nuclear facility that had been recently constructed at Windscale on the Cumbrian coast[2], including its physical placement between the sea and a background landscape of undulating mountains, which are sometimes visible. The visual similarity between the twin towers of HAMPSON's 'Interplanet Space Fleet HQ' and Windscale Piles 1 and 2 (which became operational in 1950 and 1951, respectively [HSE, 1999]), is particularly striking (see Figure 1.2).

A report on Calder Hall power station at Windscale sheds light onto some of the decisions about the visual style of this landscape:

> The architects gave much thought to the appearance and placing of the buildings. This factory was on the edge of the Lake District, one of the loveliest parts of England, and everyone con-

2 This facility began construction in 1947 as a plutonium processing plant for atomic weapons, before being developed to include Calder Hall nuclear power station in 1956.

> cerned with it was anxious that it should disturb its surroundings as little as possible ... The factory buildings are simple steel-framed structures clad in concrete and glass. No attempt is made to conceal the purpose of the factory but colour has been used in a lively way to soften its austerities (JAY 1956, 42).

This could, indeed, be a description of HAMPSON's space port, and it is notable that the modernist aesthetic of the building materials, the unity of form and function, and careful placement in the landscape are all facets shared with HAMPSON's detailed design. Moreover, the conflation of ideas implicit in HAMPSON's imagery between the promise of atomic power and the excitement of space exploration represents an interesting perspective from which to comprehend these landscapes. Indeed, the early 1950s, before the formation of the Campaign for Nuclear Disarmament in 1958, represent a period when atomic power was presented as the solution to every kind of modern-day problem (CONEKIN 2003, 68), and the atom was a symbol of modernity and a better world to come (KRIGE 2006).

Traditional landscapes of English pastoralism are occasionally portrayed by HAMPSON in addition to these idealised technological vistas. For example, one opening scene depicts a cricket match taking place on a village green, complete with church, sunshine and thatched cottages (see Figure 1.3). This expanded opening frame, incorporating the 'Dan Dare' lettering, would sit prominently alongside the *Eagle* logo on the paper's front page and set the scene for the forthcoming action, functioning in much the same way as the 'space port' frame in Figure 1.1.

This image, when considered on a standalone basis, bears resemblance to, and arguably, is heavily influenced by, other mid-century depictions of English rural landscapes such as Frank Newbould's 1940s poster series for the War Office entitled 'Your Britain – Fight for it Now' (see Figure 1.4) and Brian Cook's dust jackets from the Batsford guide books of the 1930s. Newbould and Cook both portrayed picturesque 'English village' settings in a modern style, using sections of monochrome and a primary colour set, which are similar to HAMPSON's depiction of the village green in both style and content. In their use of colour and delineation in a bold, eye-catching manner, these landscapes also share the function of publicity; effectively advertising a wartime message, a book series and a children's paper, respectively, whilst at the same time all contributing to forms of popular geographic knowledge and a certain idea of the English rural landscape (BRACE 2003). When considering this opening frame in the context of its sequence, we can see how HAMPSON develops this scenario as the setting for an alien invasion. One particularly striking frame shows the descending space ship from a low perspective, seen in the gaps between the cricket stumps, with bails still intact (HAMPSON 1951).

That the first contact on Earth for an attempted alien invasion occurs right in the middle of this idyllic setting perhaps demonstrates what HAMPSON thought might be under threat in a modernising society. More likely, however, is that HAMPSON is parodying this very notion, as represented by the reaction of the villagers, one of whose main concerns is that the alien space ships might "ruin the pitch" (HAMPSON 1951, 2). The ensuing sequence can also be seen in the context of the long-standing British invasion genre, dating back to Victorian fears of incursion through an as yet unconstructed Channel tunnel (PICK 1994), particularly with the appearance of an 'admiral' at the helm of the gathering angry mob, one of whom addresses the alien ship with the French command "*Allez!*" (HAMPSON 1951, 2). The invasion motif in science fiction can be traced back to H G Wells' *The War of the Worlds*, in which Martians start their invasion of Earth on the outskirts of Woking, and this genre had indeed been re-worked in other forms of British science fiction throughout the 1950s and 1960s (HUTCHINGS 1999). In any case, it is pertinent to echo PICK's assertion (1994, 80) that "the threat of invasion or defeat has sometimes helped to consolidate a collective sense of identity", a sentiment which is displayed through the depiction of a flag of St. George at the focal point of the scene in Figure 1.3.

HAMPSON was one of the first comics artists to employ a cinematic approach, through the use of close-ups, mid-range and long shots, as if one were positioning a camera in a film shoot. This approach represented a significant advance in comics artistry in Britain, with previous efforts, such as the aforementioned *Dandy* and *Beano*, simply treating each frame as a "stage", upon which the characters would appear (CROMPTON 1985, 29). Using this more sophisticated technique, HAMPSON was able to feature broad landscapes drawn from an aerial perspective, some of which included depictions of civic structures such as transport infrastructure and housing developments. One of the standard conventions of the comics format is the portrayal of movement through the use of motion lines (MCCLOUD 1993). They are a repeated feature of the opening pages of *Dan Dare*, used to convey the motion of flying machines over the space port landscape, as seen in the top-right corner of Figure 1.1. In a sequence which spans the opening pages of *Eagle*'s third issue, Dan is transported by "helicar", a futuristic cross between a jet plane and a helicopter, over various landscape features to an "emergency cabinet meeting" (HAMPSON 1950b, 1).

HAMPSON's depictions of these types of landscape typically incorporated modern elements into pastoral settings. Thus, in the "helicar" sequence, we see a Le Corbusier-style housing block framed in a picturesque manner by trees, and a motorway passing between ploughed fields and green spaces, with Dan flying across the sky in the background (HAMPSON 1950b, 2). Indeed, these landscapes were created at a time when modernist principles were being used in the planning, design and landscaping of large-scale projects such as the establishment of a national motorway network (MERRIMAN 2006). Some of these developments drew upon a growing sense of personal mobility in 1950s Britain, both in their production, through using new technology such as helicopters in the surveying of landscapes, and in their consumption, in the case of the motorway, through the growth of car ownership. HAMPSON's artwork directly incorporates this concept of personal mobility, as Dan skims across the page, from frame to frame, in his personal helicar. With these interpretations in mind, it can be seen how HAMPSON's domestic landscapes promote a vision of post-war British rural modernity, which consisted of industrial, civic and dynamic spaces that were intertwined with the surrounding countryside.

A major part of Dan Dare's adventures involve him exploring either outer space or other planets of the solar system. This can be seen in the context of long-standing Earth-bound narratives of exploration (Cf. DRIVER 2001), particularly when bearing in mind some scenes where Dan encounters alien civilisations. However, a dominant theme connected to this practise of exploration is that of mobility. Although a multitude of different types of transport technology are portrayed in Dan's adventures, the most important is space flight itself. When Dan begins his outer space adventure, HAMPSON explores the use of different forms of space flight machinery. The narrative opens with news of a space craft that has already been sent to Venus, which has become the latest victim of a technical fault in the "impulse cylinders", the space flight technology that has been used in Dan's world "for the past fifteen years" (HAMPSON 1950c, 1). It transpires that the only way to breach the atmosphere of Venus is to revert back technologically to the rocket motor, and thus a new rocket ship is constructed "at breakneck speed" by white-coated technicians at the Interplanet Space Fleet HQ (HAMPSON 1950c, 2). The rocket is a form of technology that HAMPSON was accustomed to from the war, and the use of rocketry for space flight would also have been discussed with Arthur C. Clarke, who was an expert on this topic, and would have been working on his non-fiction book *Interplanetary Flight* (1950) at this time. In this way Dan and his companions land on Venus, which is portrayed as a planet of lush, exotic vegetation, clearly drawing on Lewis's imaginative descriptions of the planet in *Perelandra*. However, this being an adventure story, some villains are needed, and materialise in the form of the green-skinned, reptilian Venusians known as 'Treens'. Clarke is thought to have been a significant influence on the design of the Treens and their characterisation as "scientific automatons" or "boffins run wild" (MORRIS and HALLWOOD 1998, 140), whilst HAMPSON later suggested that they were an embodiment of the Nazis (WATKINS 2000, 167), with their search for an all-powerful super-weapon and plans for the conquest of Earth.

Whilst mobility, and the technology of motion, is the key to Dan Dare's adventures in interplanetary space, this is occasionally counterpoised with a form of anti-mobility, particularly when Dan encounters his alien adversaries. Dan's weapon of choice is not a conventional side-arm; he uses a "paralysing pistol" (HAMPSON 1950d, 2) to freeze his enemies in their moment of attack. This narrative device would have served to minimise the level of violence HAMPSON put into his comics, mindful as he was of the editor's aim to create a clean-cut publication for children. However, a certain amount of menace is upheld, more through suggestion than graphic detail, in this adventure. Furthermore, when Dan is captured by the Treens at their first meeting, the antagonists use a kind of globular bubble-trap, ensnaring Dan and his companions, restricting their movement and rendering them helpless. The key point to take from HAMPSON's depiction of such weapons is that although mobility is a crucial asset for the protagonists to control, the upper hand can also be won through rendering enemies physically less mobile. Whilst this theme of mobility is borne out through such encounters in Dan Dare's universe, it is important to acknowledge the thematic connections with HAMPSON's terrestrial landscapes, and to understand these aspects of *Dan Dare* as metaphors for ways of thinking about technology, landscape and movement in 1950s Britain which were in themselves indicative of broader cultural trends.

CONCLUSION

The examples discussed in this chapter demonstrate that MORRIS and HAMPSON's "new kind of strip-cartoon paper" did not just consist of simple adventure tales illustrated in a straightforward manner, but the *Eagle* and *Dan Dare* in fact offer reflective and sometimes quite complex readings of landscape and culture at a time when British society was modernising in ways that were exciting, unpredictable and sometimes threatening. HAMPSON projects this sense of progress forward with his depictions of a future Britain and the prospect of space exploration, and by examining his "national launch pad" to interplanetary space, alongside his interpretation of the adventure hero character and its placement in the moral and technological context of the *Eagle*, it is possible to make detailed insights into representations of British post-war culture. We have briefly noted how some scholars have viewed Britain in the post-war period as a nation in decline, and how this vision has been associated with a certain Romanticised, nostalgic view of the English countryside, with claims that this type of sentimental representation of landscape had stilted a sense of progress and contributed towards what one critic has called "British retardation" (WEINER 1981, 167). However, far from encapsulating the conservative 'consensus' point of view, MORRIS and HAMPSON's *oeuvre* promotes a vision of British landscape and culture that draws upon traditional and ethical values but at the same time it is also exciting, technological and progressive. This complicates the polarised nature of the 'declinist' debate, particularly when landscape has been used to evoke certain themes, and complements more nuanced arguments which understand that the modern and traditional aspects of landscape can be held in alignment (MATLESS 1998), which can in turn help us to understand the "spatial fabric of the modern" (GILBERT et al. 2003) in post-war British culture. It is hoped that this chapter, as well as providing a geographical perspective on this debate in cultural history, has also contributed towards an understanding of comics history in Britain, and has provided some modest insights into the creation of one of Britain's most fondly-remembered cultural institutions.

REFERENCES

ADDISON, P. (1985): *Now the War is Over – A Social History of Britain, 1945–1951.* London.

BARKER, M. (1984): *A Haunt of Fears – The Strange History of the British Horror Comics Campaign.* London.

BARNETT, C. (1986): *The Audit of War: The Illusion and Reality of Britain as a Great Nation.* London.

BOWLER, P. (2006): Experts and Publishers: writing popular science in early-twentieth century Britain, writing popular history of science now. *British Journal for the History of Science* 39 (2), 159–187.

BRACE, C. (2003): Envisioning England: the visual in countryside writing in the 1930s and 1940s. *Landscape Research* 28 (4), 365–382.

CHAPMAN, J. (2011): *British Comics – A Cultural History.* London.

CONEKIN, B. (2003): *'The autobiography of a nation' – The 1951 Festival of Britain.* Manchester.

CROMPTON, A. (1985): *The Man Who Drew Tomorrow.* Bournemouth.

DANIELS, S. and D. COSGROVE (1988) Iconography and Landscape. S. DANIELS and D. COSGROVE (Eds.): *The Iconography of Landscape: essays on the symbolic representation, design and use of past environments.* Cambridge, 1–10.

DITTMER, J. (2005): Captain America's Empire: Reflections on identity, popular culture, and post-9/11 geopolitics. *Annals of the Association of American Geographers* 95 (3), 626–643.

DOWNING, D. (1992): *Planets in Peril – A Critical Study of C S Lewis's Ransom Trilogy.* Amherst.

DRIVER, F. (2001): *Geography Militant: Cultures of Exploration and Empire.* Oxford.

DUNNETT, O. (2009): Identity and Geopolitics in Hergé's Adventures of Tintin. *Social and Cultural Geography* 10 (5), 583–598.

EDGERTON, D. (1991): *England and the Aeroplane – An Essay on a Militant and Technological Nation.* London.

EDWARD, J. (1987): The Future Viewed From Mid-Century Britain: Clarke, Hampson and the Festival of Britain. *Foundation – The Review of Science Fiction* 41, 42–51.

EISNER, W. (1985): *Comics and Sequential Art.* Taramac, Florida.

GILBERT, D., D. MATLESS and B. SHORT (2003): Historical Geographies of British Modernity. D. GILBERT, D. MATLESS and B. SHORT (Eds.): *Geographies of British Modernity.* Oxford, 1–28.

GRAVETT, P. and P. STANBURY (2006): *Great British Comics – Celebrating a Century of Ripping Yarns and Wizard Wheezes.* London.

HAMPSON, F. (1950a): Dan Dare. *Eagle* 14/4/50, 1–2.

HAMPSON, F. (1950b): Dan Dare. *Eagle* 28/4/50, 1–2.

HAMPSON, F. (1950c): Dan Dare. *Eagle* 5/5/50, 1–2.

HAMPSON, F. (1950d): Dan Dare. *Eagle* 9/6/50, 1–2.

HAMPSON, F. (1951): Dan Dare. *Eagle* 10/8/51, 1–2.

Health and Safety Executive (HSE) (1999): Nuclear Reactors (Environmental Impact Assessment for Decommissioning) Regulations. http://www.hse.gov.uk/nuclear/windscale.pdf (accessed May 2, 2012).

HEWISON, R. (1997): *Culture and Consensus – England, Art and Politics since 1940.* London.

HOGGART, R. (1957): *The Uses of Literacy.* London.

HUTCHINGS, P. (1999): 'We're all Martians now' – British SF Invasion Fantasies of the 1950s and 1960s. I. Q. HUNTER (Ed.): *British Science Fiction Cinema.* London, 33–47.

JAY, K. (1956): *Calder Hall – The Story of Britain's First Atomic Power Station.* London.

KRIGE, J. (2006): Atoms for Peace, Scientific Internationalism and Scientific Intelligence. *Osiris* 21, 161–181.

KYNASTON, D. (2007): *Austerity Britain – 1945–51.* London.

MATLESS, D. (1998): *Landscape and Englishness.* London.

MCALEER, N. (1992): *Odyssey: The Authorised Biography of Arthur C. Clarke.* London.

MCCLOUD, S. (1993): *Understanding Comics: The Invisible Art.* New York.

MERRIMAN, P. (2003): 'A Power for Good or Evil': Geographies of the M1 in Late Fifties Britain. D. GILBERT, D. MATLESS and B. SHORT (Eds.): *Geographies of British Modernity.* Oxford, 115–131.

MONTGOMERY, J. (1965): *The Fifties.* Leicester.

MORRIS, M. (1949): 'Comics' That Take Horror into the Nursery. *Sunday Dispatch*, 13/2/1949, 1.

MORRIS, M. (1950): Editorial. *Eagle* 14/4/50, 3.

MORRIS, M. (1977): Introduction. D. GIFFORD (Ed.): *The Best of 'Eagle' Annual, 1951–59*. London, 3–17.

MORRIS, S. and J. HALLWOOD (1998): *Living With Eagles: Marcus Morris, Priest and Publisher*. Cambridge.

National Archives (2012) 'The Art of War' online exhibition. http://www.nationalarchives.gov.uk/theartofwar (accessed June 1, 2012).

PICK, D. (1994): Pro Patria: Blocking the Tunnel. *Ecumene* 1(1), 77–93.

SHARP, J. (2000): *Condensing the Cold War: Reader's Digest and American Identity*. Minneapolis.

SILLARS, S. (1995): *Visualisation in Popular Fiction, 1860–1960*. London.

STABLEFORD, B. (1986): Fantasy. P. NICHOLLS (Ed.): *The Encyclopedia of Science Fiction*. London, 210–212.

VARAH, C. (2004): Morris, Marcus Harston (1915–1989). *Oxford Dictionary of National Biography*. http://www.oxforddnb.com/view/article/40164 (accessed July 26, 2010).

VELDMAN, M. (1994): *Fantasy, the Bomb and the Greening of Britain – Romantic Protest, 1945–1980*. Cambridge.

VINCE, A. (2007): *Letter to Eagle Times* 21(3), 37.

WATKINS, T. (2000): Piloting the Nation – Dan Dare and the 1950s. D. JONES and T. WATKINS (Eds.): *A Necessary Fantasy? The Heroic Figure in Children's Popular Culture*. London, 153–176.

WEIGHT, R. (2002): *Patriots – National Identity in Britain 1940–2000*. London.

WEINER, M. (1981): *English culture and the decline of the industrial spirit 1850–1980*. Cambridge.

WILLIS, C. (1950): The Fires Within – A Story of the Future. *Eagle* 4/10/50, 12–13.

10TH APRIL, 1999, CONWAY HALL, RED LION SQUARE: *SNAKES AND LADDERS*, OCCULT CARTOGRAPHY AND RADICAL NOSTALGIA

> [I]t is utterly true that he who cannot find wonder, mystery and awe, the sense of a new world and an undiscovered realm in the places by the Gray's Inn Road will never find those secrets elsewhere. All wonders [...] lie within a stone's throw of King's Cross Station.
> (Arthur MACHEN 1923, 323)

> I'll tell you the ultimate secret of magic. Any cunt could do it.
> (Alan MOORE 2001, n.p)

INTRODUCTION

10th April, 1999, Conway Hall, Red Lion Square. The one off mixed media performance of *Snakes and Ladders* took place featuring free form spoken word recital from Northampton-based comics writer and self-proclaimed magician, Alan MOORE. This was one of a series of collaborative performance pieces that Moore continues to be involved with as part of the grandly titled Moon and Serpent Grand Egyptian Theatre of Marvels; a collective with a revolving membership. The evening performance of *Snakes and Ladders* was the climax to a conference on magic, which a review called a "symposium of real magick and global ritualism [...] hosted by the Oxford Golden Dawn Occult Society," adding that the conference addressed itself to "Golden Dawn or Crowley people, sethians, hyperdelics, graphic novelists, voodooists, modern primitives or post porn modernists" (DARK 1999, n.p.). The performance itself featured music from regular MOORE collaborator Tim PERKINS, as well as a light show, candles, incense, fire breathers, living statues, and dancer Paula van Wijngardeen with a puppet snake. MOORE's contribution was a poetic verbal excavation of the square's history featuring guest appearances from (among others) Oliver Cromwell, Dante Gabriel Rossetti and Lizzie Siddal, and Francis Crick. The closest thing to a connecting narrative thread can be found in the grieving figure of Arthur MACHEN, a former resident of nearby number 4 Verulam Buildings in Gray's Inn and adept of the original Order of the Golden Dawn, drifting along the deliriously altered lunar landscape of the Gray's Inn Road on an allegorical journey through darkness into light. To quote briefly: "Our various threads converging with the history of this location, braided into a macramé of event unique and unrepeatable in space-time" (MOORE 2001, from Part I 'The Gate of Tears', n.p.). Conway Hall was thus transformed into a heterotopic stage connecting an assemblage of disparate personages and moments.

Conway Hall itself is a Grade II listed building in the Holborn district, between Theobald's Road and Lamb's Conduit Passage, the main entrance opening out onto the north east corner of the square. The hall is home to the South Place Ethical Society, a liberal humanist organisation with roots stretching back to the eighteenth century.[1] In 1661, the bodies of the Parliamentarian regicides Oliver Cromwell, John Bradshaw, and Henry Ireton were exhumed (in the case of Cromwell, the former Lord Protector, from Westminster Abbey) for desecration and posthumous execution at Tyburn followed by burial in a pit in what is now Red Lion Square.[2] The square was laid out on a seventeen acre rural paddock in 1684 by architect Nicholas Barbon. The lawyers at nearby Gray's Inn objected to losing their uninterrupted rural vistas and a riot erupted with the workmen, who, led by Barbon, won. Barbon, a speculative developer and son of an Anabaptist parliamentarian, was responsible for developing much of the area around High Holborn.[3] In the nineteenth century Dante Gabriel Rossetti lived at number 17, as did, later, William Morris and Edward Burne-Jones. The square's history is woven into MOORE's prose incantation, rendering contemporal different periods and personalities.

MOORE's words were subsequently adapted into comics form by artist Eddie CAMPBELL, remediating MOORE's verbal assemblage of references, allusions and historical detail into a visual-textual collage. CAMPBELL's delirious pages comprise a spatial-historical archive of sorts, incorporating great drifts of materials – photographs, classic paintings, verbal quotations and so on – alongside his own distinctive drawings. The design reinforces the correlation between space and visuality, pointing to how space and place are structured by the visual field. Visuality, according to Hal FOSTER, refers to "how we see, how we are able, or allowed, or made to see, and we see this seeing and the unseeing there in" (FOSTER 1988, ix). *Snakes and Ladders* works to make visible what has been occulted, obscured, or discarded with regards to a specific place.

Both performance and comic are examples of MOORE's excursions into London psychogeography, evidence of a geographical imagination[4] at work in MOORE's writing. Both emphasise the tropes of theatrical performance and the

1 See MACKILLOP 1986, 41-81, for a concise history.

2 For a condensed history of the square see WEINREB et al. 2008, 684.

3 "In some ways Barbon was the archetype of the modern property developer. He built on a large scale, well or poorly as the occasion demanded, used standardised designs and fittings, always kept a close eye on the market, destroyed ancient mansions and playgrounds without remorse, bullied and bluffed those whose property stood in his way, sweet-talked or evaded his clients and creditors, and died with his debts unpaid" (INWOOD 2000, 256).

4 The term "geographical imagination" stems from David HARVEY's use to describe and convey the importance of spatial locations in social and cultural arenas (HARVEY 1973, 24-27). In the field of visual studies the term is used by Joan M. SCHWARTZ and James R. RYAN as "a way of thinking" that "alerts us to the significance of the spatial in all aspects of social existence" (SCHWARTZ and RYAN 2003, 6).

reflexive performativity of signification.[5] Both articulate in different ways the affective web of associations that stretch out singularities of time and place, and have become associated with the cultural products and practices of psychogeography. Despite the growing body of academic criticism on MOORE, his numerous psychogeographical works have been comparatively overlooked, or abstracted from the contexts of British psychogeography.[6] Consequently, the often elided over aspect of nostalgia embedded within psychogeography has not been commented on in relation to MOORE's work. This essay will map a trajectory from *Snakes and Ladders* through the occult mapping of London, via the comics semiotics of Thierry GROENSTEEN, to the further reaches of British psychogeography and the structure of feeling that social geographer Alastair BONNETT calls radical nostalgia.[7]

Snakes and Ladders

Snakes and Ladders was not MOORE's first involvement with psychogeography and multi-media performance. The Moon and Serpent Grand Egyptian Theatre of Marvels had been invited by the Oxford Golden Dawn Occult Society, which itself had claimed some descent from the original Hermetic Order of the Golden Dawn. More commonly known simply as the Golden Dawn, the society was a self-proclaimed magical order active largely during the late nineteenth and early twentieth centuries and organised around syncretic pagan beliefs and practices (DAVIES 2012, 101–103). Among its illustrious members were the notorious occultist Aleister Crowley, and writers Arthur MACHEN and WB Yeats. The order declined in the inter-war years, and withered away in the post-war period. The Oxford Golden Dawn Occult Society was one of many groups which asserted a

5 I am differentiating between performance and performativity here. The former I use in relation to the actual performance at Conway Hall and the subsequent graphic theatricalised tropes that CAMPBELL uses (stage, dancers, playbills etc.); the latter as a speech act as defined by J.L. AUSTIN, who differentiated between performative and constative utterances, the former referring to an act that does something as well as saying it (AUSTIN 1962). In other words, the performative statement is a constitutive act. More specifically, I am using it here to describe CAMPBELL's *bricolage* comics design which explicitly foregrounds its constructedness, the form becoming the content as it were.

6 Annalisa DI LIDDO's reading of *Snakes and Ladders,* for instance, refers to psychogeography in passing without acknowledging Parisian antecedents (DI LIDDO 2009, 326).

7 BONNETT does not use the term structure of feeling, which I have appropriated from Raymond WILLIAMS. For WILLIAMS a structure of feeling is a "cultural hypothesis" for "social experiences *in solution* as distinct from other social semantic formations which have been *precipitated*" (WILLIAMS 1977, 133–134). For WILLIAMS, this indicated a set of flexible but interlocked beliefs and affects particularly associated with emergent cultural formations. I am using it here to indicate the residual qualities of psychogeography, which retain a sense of flux and ambiguity.

genealogical link to the original order pre- and post- millennium.[8] MOORE's collaborative work with The Moon and Serpent Grand Egyptian Theatre of Marvels stemmed from his "magical turn" on his fortieth birthday when he announced to friends and family that he was going to become a magician and so set about immersing himself in lore, ritual, and psychoactive drugs (MILLIDGE 2011, 248–251).

Though he worked with the author's permission, CAMPBELL adapted the text independently working without one of MOORE's notoriously detailed and elaborate scripts.[9] The result does not resemble most of MOORE's better known comics. CAMPBELL largely dispensed with the rigid, rectilinear variations on the six or nine panel grids so favoured by MOORE elsewhere, and instead attempted to recast the text within a mutating imagistic landscape that acted as a corollary to the performance. CAMPBELL's spidery scrawl inserts MOORE's words into a visual archive; the emphatic materiality of the page constantly adverted by the accumulated maps, photographs, paintings, portraits and of course, CAMPBELL's own pictures. Visual quotations reference celebrity photographs, pre-Raphaelite art, and Lady Frieda Harris's tarot designs. The allusive, reflexive text is matched by the piling up of fragmented images that generate a sense of polytemporal tension; an uncovering and re-aligning of the Square's history, and that of the surrounding area. Taking the Conway Hall performance as a nodal starting point, *Snakes and Ladders* the comic appropriately spirals off into a perambulant meditation on space, time, and the affectivities of place. To get a sense of this let us turn to an excerpt from part one of *Snakes and Ladders*, "The Gate of Tears" (Figure 2.1).

The double-page spread stretches the definition of what constitutes a comic. The first impression is one of confusion; where does one commence reading? Western convention guides us through practice to read from top left, across and down. The eye can also follow the text of MOORE's words which crawl slightly more haphazardly across the page, framing the illustrations. The images themselves incorporate reproductions of a photograph, a painted portrait, a map, hazy sketches of Conway Hall and various conference delegates.

Comics theorist Thierry GROENSTEEN's model of comics as a visual language is useful in considering the sheer density of these pages (GROENSTEEN 2009). GROENSTEEN proposes an architectonic semiotic system of studying comics as an interdependent network of images that function to constitute articulation. This articulation operates at a spatio-topical level – that is through the spatial/locative layout of pages – and then via sequential and disseminated associations and connections; the relationship between images that allow meaning to be produced. GROENSTEEN is interested in the relations between composite elements, especially the frames of aggregated panels, the hyperframe of the page, and the multiframe of the physical text.

---

8 The London lodge of the Oxford Golden Dawn Occult Society was mainly active between 1996 and 2000. An online archive can be found at this link: http://www.lawbright.com/logdos/ (accessed 14 July 2012).

9 See for example MOORE's script chart for *Big Numbers* in MILLIDGE 2011, 170-173.

GROENSTEEN uses the neologism "arthrology" to refer to the different types of articulations that occur between panels. "Every panel exists, potentially if not actually, in relations with each of the others" (GROENSTEEN 2009, 40). These relations are linear and sequential – restrained arthrology as GROENSTEEN puts it; and also non-linear and discursive – general arthrology. He describes 'braiding' (*tressage*) as the principle for making local and global connections in a comic. Braiding reveals the potential relations between panels that may be physically and contextually separated but which nevertheless communicate which other through, for instance, the repetition of iconic content. Rather than reducing comics to a hypothetical minimal signifying unit, GROENSTEEN focuses on the compositional relationality of component elements in a comic. Starting with the panel and its frame GROENSTEEN analyses successive linear and translinear relations.

Returning to Figure 2.1, we can see that CAMPBELL's visual strategy is to incorporate and break down the various framing devices that comics make use of. Although we can discern panel shapes, the dominant framing device is the hyperframe of the two pages. "The hyperframe" writes GROENSTEEN, "is to the page what the frame is to the panel" (2009, 30). The double page spread here is a unit in itself. The use of the image of Oliver Cromwell is doubled to emphasise iconic solidarity – his figure dominates the right hand side of the left page, and his visage is repeated and mirrored on the bottom left of the right hand page. The right hand page also includes misaligned lines that indicate a skewed page within the material page itself, the framing lines stretching over onto the bottom of the left hand page.

GROENSTEEN places a special significance on the importance of margins as peripheral zones. The margin is "not required to be empty; it is not forbidden to forms of representation" (GROENSTEEN 2009, 31). Marginal space is not therefore just a base on which to overlay a grid of images but may influence the articulation of the system. In this case, the background is unevenly white. While we can see that there is a line of vertical white space on the right edge of the right hand page that constitutes what we would normally define as a margin, much of the rest of the pages occupy an indistinct space that foregoes zonal separation. There is a ghostly impression submerged underneath the pages – the map on the left-hand page fades unevenly, while the sketch of Conway Hall is overlaid onto a barely perceptible photographic image. The material qualities of the images are themselves emphasised: as well as the portraits of Cromwell, seemingly ripped from other sources and recontextualised, we see a drawing of scientist Francis Crick[10], and a black and white photograph of *fin de siècle* writer Arthur MACHEN (associated briefly with the Decadents, and better known for his gothic horror stories). The picture of Crick, the photograph of MACHEN, and the oval shaped portrait of

10 Most famous for his work with James D. Watson on uncovering the structure of DNA – Crick's tie, stabbing out of the image for emphatic resonance, contains a spiral pattern which as well as acting as a reference to DNA provides an instance of iconic solidarity pointing to both further serpentine visual patterns as well as the twisting performative structure of the comic.

Cromwell all constitute frames for the associated images. Such frames, argues GROENSTEEN, are "at the same time the trace and measure of the space inhabited by the image" (GROENSTEEN 2009, 29). The images of MACHEN and Cromwell in particular are examples of what GROENSTEEN calls "prographic" materials whereby the panel "takes up or integrates, with or without modification, an earlier, generally photographic document" which results in a "second degree utterance, an image of the image, a citation" (GROENSTEEN 2009, 41–42).

GROENSTEEN consistently argues that comics are a predominantly visual medium, but acknowledges the importance of the linguistic and its interplay with the iconic. The written text commences in the top right hand corner, comes down as a column, then across the middle of the page, recommencing once more on the left hand page where it is patterned as three horizontal strips. The effect is one of anchorage and relay of the images, the text performing a "rhythmic function" of controlling the pace of the reading (GROENSTEEN 2009, 131–133). "Our stories, our back-continuities that herded us down separate tracks toward this room" – the lines clearly announce a thematic interest in the historical and the temporal linked to the space of performance and continually supplemented by the images.

It should be said that GROENSTEEN by his own admission ignores any consideration of comics as "historical, sociological, or economic phenomena" (2). GROENSTEEN's spatio-topial model identifies the elements that constitute the material space of the comics page, but requires extra-textual components to adequately address the affectivities that inhere in the representations of the physical and temporal attributes of space and place. For instance, while recognizing the usefulness of GROENSTEEN's quasi-structuralist approach, geographer Jason DITTMER has pointed out that despite trumpeting "plurivectoral narration" GROENSTEEN's "insights tend to limit the narrative to a singular thread that readers have some role in constructing. For example [...] there is one 'correct' ordering of the panels, even if you read back and forth along that ordering" (DITTMER 2010, 231). DITTMER is keen to stress the importance of montage in comics – a term GROENSTEEN specifically refutes as he associates it with the cinema and the post-production practices of editing linear filmic shots (GROENSTEEN 2009, 101–102). But this is to ignore the aesthetic and political affinities that montage shares with modernist and postmodernist practices of collage and bricolage, and particularly how they relate to psychogeography.

These practices provide the methodological motor for Walter BENJAMIN's crucial work on urban history, itself something of an *ur*-text for psychogeographical thought and work. BENJAMIN's monumental *Arcades Project* attempted to write a history of capitalism through the prism of the Paris arcades of the nineteenth century. BENJAMIN's method was to "develop to the highest degree the art of citing without quotation marks. Its theory is intimately related to that of montage" (BENJAMIN, N1, 10). BENJAMIN's *Arcades* embodies in performative form an assemblage of quotations and commentary whose principle of construction anticipates psychogeography. The next section will take a detour, in the footsteps of Arthur MACHEN and others, following the well worn pathways of psychogeography, from Paris to London and beyond.

IN THE TRACKS OF ARTHUR MACHEN

Unsure where to start *Snakes and Ladders*, MOORE goes back to the Big Bang, returning swiftly to present day Red Lion Square; the visual leitmotif of the double helix threads its way throughout. CAMPBELL follows MOORE's structure of five parts, moving digressively from Conway Hall through to meditations on art, performance, and place. The five sections are also named after the songs that made up the performance and make up the subsequent CD. Following "The Gate of Tears", "Stars and Garters" opens with a mock programme for a Victorian music hall extravaganza involving a female dancer and a snake as the personification of the principle of the universe according to Frieda Harris' tarot designs; the third section entitled "Baghdad" follows the pitiful figure of MACHEN as he navigates a nightmarish landscape projected by his depression at the death of his wife; the fourth section, "Art", is the shortest but possibly the most important in providing a performative reflection on the transformative power of the imagination; concluding, 'Syon' brings these discursive cultural, historical, and geographical movements to an enraptured conclusion as the mourning MACHEN walks back into the light and into the sunshine of Red Lion Square. Spiraling through this comic antinarrative the forlorn figure of MACHEN acts as a navigator for the reader, lending a sense of visual and semantic continguity. MACHEN's loss figuratively deforms the landscape of the Gray's Inn Road, turning it into the nightmarish Baghdad, an imaginary backdrop to his intense melancholy (Figure 2.2).

Once again, CAMPBELL uses the double-page spread as a hyperframe, the distorted London skyline dominating the vision of the reader (see Figure 2.2). Pursued by his own personal demons, MACHEN is chased by characters from his novel *The Three Imposters*. Although there are the occasional orientalist flourishes in text and image, this is not a reference to the actual Middle Eastern city but rather "Baghdad is the name Machen gave to the imaginary lunar landscape he claimed to have visited in the nightmarish state of mind that followed his wife's death" (DI LIDDO 2009, 330).[11]

The capital here is transformed into a phantasmagorical landscape, dwarfing the figure of MACHEN and his threatening, unreal shadows. The lopsided and cluttered perspective collapse time and space into the hyperframe. The inky blackness of the river is foregrounded, while disappearing into the tangled city. In the left hand corner of the left hand page, a black bird oddly out of proportion arches its wings eerily resembling a statue come to life underscoring the gothic ambience. The image fills the hyperframe to the edges of the pages, obliterating any potential margins. MACHEN appears to be running from the illuminated white space of the streets at the bottom of the right hand page, toward the bridge and over the river toward uncertain darkness.

---

11 It was following the death of his wife that MACHEN drifted toward the Order of the Golden Dawn, one of many occult societies that flourished during the *fin de siècle* (see VALENTINE 1995, 72, 75).

This movement, from right to left, coincidentally runs counter to Western reading conventions which follow text left to right and once again means that the reader must pause to fully comprehend the page. The "textual zones" themselves "compose a network in which the general configuration is a determining factor in the protocol of reading" (GROENSTEEN 2009, 69). The text is contained in black boxes, white text on a funereal black background adding another gothic flourish.

A crucial influence both on MOORE and London psychogeographical writing in general, for MACHEN, the exotic and esoteric was located in the everyday "beneath the crust of common and commonplace things; hidden and yet burning and glowing if you care to look with purged eyes" (MACHEN 1924, 75) MACHEN's own late Victorian gothic fiction obsesses over borderlines and boundaries. In the case of his most notorious work, the novella *The Great God Pan* (1894), the colonial margins on the Celtic fringe of the wild Welsh border are folded into the rational, urban setting of London as imperial metropolis. The membrane between the ordered and the sensual is breached and the uncanny presence of the pagan god is felt in the capital. This is a classic movement in colonial gothic tracing a trajectory from the colonial margins to the imperial centre, but also hinting at a further passage between incommensurable parallel spaces.

PSYCHOGEOGRAPHIC GENEALOGIES

MOORE has made numerous contributions to the sub-genre of British psychogeography[12], a term famously categorised with "pleasing vagueness" by Guy DEBORD to indicate the affective intensities of predominantly urban spaces and attempts to resist the disciplined instrumentality of such spaces (DEBORD 1955, 5). This influential, if gloriously imprecise, definition was an attempt to transform an increasingly ordered and instrumentalised space for aesthetic and political ends. From Paris, and an association with post-war continental left-wing groupings, psychogeography crossed the channel and inspired what Merlin COVERLEY has called "retrospective validation in traditions that pre-date DEBORD's official conception by several centuries" (COVERLEY 2006, 11).[13] This retconning of the term has

12 It may be that MOORE's greatest contribution to psychogeography is his Northampton set writing, especially in his formally experimental historiographic metafiction *Voice of the Fire* (1996) and the sadly incomplete comics collaboration with Bill SIENKIEWICZ *Big Numbers* (1990), not to mention the long anticipated second novel *Jerusalem.* The links to Northampton are spelled out in the narrative text of *Snakes and Ladders* in figure 3.1 which positions Red Lion Square and Holborn as part of a geographical and historical network connected to MOORE's home town: "From this complex lace of moment, all our strands of individual information stretch away towards their different points of origin, through how we got here, winding back to where we're from. In my case, that would be Northampton, an unprepossessing blur on the M1 halfway to Birmingham" (n.p.)

13 COVERLEY points out that DEBORD himself was "dismissive of attempts to establish psychogeography within the context of earlier explorations of the city" (10). Nevertheless, this has not stopped people doing just that especially in English writing.

arisen largely in late twentieth century English literature and film, and includes such diverse practitioners as Iain SINCLAIR, Peter Ackroyd, and Stewart HOME, as well as a retrohistoricised genealogy of (largely gothic) London writing that ropes in Blake, De Quincey, Stevenson and, of course, MACHEN. Alongside this visionary literary trend has been the growth of localised psychogeography groups as part of a more politicised radical subculture continuing the tradition of DEBORD et al. Psychogeographical groups were established in Manchester, Nottingham and North East Essex as well as London. At the same time, groups started up in New York and Washington, and Bologna and Milan. Social geographer Alaistair BONNETT notes that from "the late 1990s activity among these revolutionary groups diminished and interest in psychogeography passed to the arts community" (BONNETT 2009, 65 n2; 59).[14] Both literary and political approaches were concerned with notions of re-mapping the city; a point echoed by writer and film-maker Iain SINCLAIR's call for an "occult mapping" to re-awaken London's "dormant energies" (SINCLAIR 1997, 1)

MOORE's understanding of psychogeography has been influenced by SINCLAIR, his friend, contemporary, and sometime collaborator. Perhaps the best known contribution to the field is MOORE's most famous collaboration with CAMPBELL, the graphic novel *From Hell*: a metaphysical inquiry into the infamous Whitechapel murders of 1888 that doubles as a reflexive commentary on the dubious publishing phenomenon of "Ripperology", complete with its own comprehensive appendices. In chapter 4, William Gull and John Netley attempt a Masonic reterritorialisation of the city, inspired by SINCLAIR's *Lud Heat* and the occult energies of Hawksmoor's churches. More recently MOORE has contributed a prose piece, "Unearthing", to SINCLAIR's anthology, *London: City of Disappearances*; which, following the reverse trajectory of *Snakes and Ladders*, has been adapted into a performance piece with music and then recorded and released on limited edition vinyl and CD. "Unearthing" is ostensibly a biography of his friend and fellow comics writer, Steve Moore (no relation), and mutates into a geographical and historical "unearthing" of Shooters Hill in south London. In *The League of Extraordinary Gentlemen: Volume III*, in which every character has been textually poached from prior genre fictions and stitched into a recombinant narrative, SINCLAIR himself even manages to put in a personal appearance as Andrew Norton, the prisoner of London from *Slow Chocolate Autopsy,* a figure trapped in the space of London but able to move across time.[15] Most recently MOORE collaborat-

14 BONNETT has some direct connection to these groupings having edited *Transgressions: A Journal of Urban Exploration* between 1995 and 2000.

15 *Slow Chocolate Autopsy* was a collaboration between SINCLAIR and graphic designer/comics artist Dave McKean. Prose chapters alternated with "graphic stories" which foregrounded montage-as-method suggesting a single unit fractured into shards creating a "London *Passagenwerk.*" Similarities with *Snakes and Ladders* are obvious. See VENEZIA (2011): "Instead of establishing causal connections, this comics montage rearranges and juxtaposes dialectical image and text producing profane illumination" (n.p.).

ed with SINCLAIR and others for *An English Journey Re-imagined* at the Snape Maltings in Aldeburgh; a revising of J.B. Priestley's 1934 book.

The most remarkable psychogeographical contributions remain the site specific performances that MOORE has undertaken with the Moon and Serpent collective. *The Birth Caul: A Shamanism of Childhood* was performed at the Old County Court in Newcastle-upon-Tyne on November 18th 1995. The piece presented a reverse chronology of an author analogue; through adulthood, to infancy, to a prenatal state. The piece was also adapted into a comic by CAMPBELL. More pertinently, *The Highbury Working: A Beat Séance* was performed on November 20th 1997 at the Highbury Garage and invoked former denizens of the area including once again the Great Beast himself, Aleister Crowley, troubled record producer Joe Meek, and Romantic poet Samuel Taylor Coleridge, all in the shadow of the former Arsenal football stadium. All of these pieces highlight MOORE's fascination with magic, and decisively link the occult to performance and psychogeography.

These projects all speak to an attempt to retrieve marginal spaces and histories, and illustrate psychogeography's work that "explores and re-imagines the forgotten nooks and crannies of ordinary landscapes. It seeks to re-enchant and re-mythologise prosaic geographies" (BONNETT 2009, 46). As unofficial psychogeographer-in-chief, SINCLAIR provides a link between on the one hand a series of proliferating collaborative projects across media, and on the other with a set of practices derived and adapted from the Paris based Situationists tempered with both the influence of the counter-culture and a longer tradition of British neo-Romanticism.[16]

DEBORD had talked of psychogeography as a "renovated cartography": "The production of psychogeographic maps, or even the introduction of alterations such as more or less arbitrarily transposing two maps of two different regions, can contribute to clarifying certain wandering that express not subordination to randomness but insubordination" (DEBORD 1955, 7). The purpose was to encourage the art of the *dérive* or the purposeful drift as a form of willed disorientation, forging new routes and passages in an attempt to unleash the emotional intensities of urban spaces. By dismantling and recomposing maps, psychogeographers sought to defamiliarise the over familiar, a refined bewilderment of the senses that challenged rationalist perspectival grids. As an example of this, DEBORD and fellow psychogeographer Asper Jorn created an absurdist collaged cartography in *The Naked City* by cutting out popular Parisian landmarks from a well-known tourist map and rearranging them on a blank page. These landmarks were then re-connected by a series of arrows, joining areas of the city separated by brute material reality in imaginary contiguity.[17]

The collage mapping was an act of *détournement*, a derailing or detouring of meaning by the appropriation and defacing of already existing documents. Writ-

---

16 For an overview of British neo-Romanticism in the twentieth century see WOODCOCK.

17 The map is reproduced in MCDONOUGH 2009, 89.

ing in "A User's Guide to Détournement" with Gil WOLMAN, DEBORD defines a tactic whereby "[a]ny elements, no matter where they are taken from, can be used to make new combinations" (DEBORD and WOLMAN, 15). As with their practices of urban wandering which succeeded Surrealist *flânerie, détournement* was a continuation of Dada and Surrealist practices of collage-as-juxtaposition.

Psychogeographic collage resonates with Walter BENJAMIN's study of another period of modernity: "To write history thus means to cite history. It belongs to the concept of the citation, however, that the historical object in each case is torn from its context" (BENJAMIN 1999, N11, 3). For BENJAMIN, this method constituted a constellative critical method for grasping the relations between historical moments, a process of preservation and recovery that brings together and reveals the past as recognised within the present, a "re-constellation of lost objects of the everyday" (BUSE et al., 33). BENJAMIN's historiography suggested that reality itself had been rendered phantasmagorical by the processes of commodity fetishism that emerged in capitalism after the French revolution. Rather than proposing a hermeneutical ideology critique, BENJAMIN suggested re-enchantment offers possibilities by re-circuiting the distortions and disorientations of the market place:

> One may compare the pure magic of those walls of mirrors which we know from feudal times with the oppressive magic worked by the alluring mirror-walls of the arcades, which invite us into seductive bazaars (BENJAMIN 1999, R2a, 2).

This seems to be suggesting that magic is not mere illusion, but can be put to work as "counter-magic" against the lure of the market place and the celebration of history as a story of linear progress.[18] Writing on the Surrealists, BENJAMIN proposed that a "profane illumination" was possible to overcome religious illumination (in this case the Catholicism of many French Surrealists). Such illumination can be found by a derangement of the senses by the use of hashish and opium, but also finds a corollary in Surrealist novels, such as Louis Aragon's *Paris Peasant* and Andre Breton's *Nadja* and their delirious ambulatory descriptions of Paris (BENJAMIN 1999a, 208–209). This sense of spatial derangement and disarrangement also connects with MOORE's take on magic and the influence of the Arts Lab and the counter-cultural counter-conjurations of London psychogeographers.

MAGICAL TURNS

MOORE's take on magic is fundamentally cultural, following Crowley's notion of magic as a disease of language: this is the idea of magic not so much as an art, but

18 See also COHEN, who analyses and assesses what she terms BENJAMIN and Breton's "Gothic Marxism: "[I]f images of the past spring to legibility in the present, it is because they speak to its concerns" the most suggestive being "the valorisation of the realm of a culture's ghosts and phantasms as a significant and rich field of social production [and] the valorisation of a culture's detritus and trivia as well as it's strange and marginal practices" (1993, 11).

as art itself (MILLIDGE 2011, 248–251). This explicitly performative aspect of magic chimes with MOORE's experience working with the Northampton Arts Lab in the early 1970s (see DI LIDDO 2009, 327).[19] This experimental open space provided access to poetry readings, theatre, and multi-media installations. The linking of mixed media performance to the practice of magic is taken up by Simon DURING in his book *Modern Enchantments*. DURING maps out a cultural history of what he calls "secular magic", i.e. the magic of conjuring shows and technical special effects. DURING makes the point that secular magic can never entirely be "disentangled from its opposing twin, magic with a supernatural punch" (DURING 2002, 2). Nevertheless, from the Enlightenment onwards, the secular gradually marginalises the sacred. In examining the place of secular magic as a modern cultural formation, DURING theorises the "magic assemblage":

> That motley of shows in the public spaces where magic was performed: theatres, fairs, streets, taverns, and so on [...] At one time or other, the magic assemblage included optical illusions like magic lantern shows or early film, feats of strength, juggling, posture mastery, ventriloquism, puppet shows, trained animal exhibitions, comics routines, automata displays, lotteries, and joy rides (DURING 2002, 66).

In other words, magic is something like a mixed media presentation from the Arts Lab or the Moon and Serpent collective. There is however, one key difference: as DURING notes, secular magic developed historically as part of the leisure industries. Audiences were attuned to view illusion *as* illusion. In contrast, MOORE's inclusive conception of magic is linked to consciousness raising. MOORE presents a magically inflected psychogeography; playful and nostalgic, informed by his professed anarchist politics. Woven into *Snakes and Ladders* is a brief encounter between MOORE and his fictional creation, working-class sorcerer John Constantine.[20] Constantine tells him the secret of magic is that "any cunt could do it." This moment of creator meeting character is a repetition of MACHEN's encounter with the three imposters. The difference is that rather than being sinister, this meeting is more affirmative and articulates a democratic principle in association with the occult. Magic becomes a process to attempt to re-enchant the banal everyday. Such re-enchantment places an emphasis on spaces of transformation.

Transitional states were important to BENJAMIN in his own attempts at envisioning a mythological topography. Modernity has left us impoverished. The traditional threshold experiences – "the ceremonies that attach to death and birth, to marriage, puberty, and so forth" – have declined (BENJAMIN 1999, O2a, 1). The collage magic of *The Arcades Project* can be read as an attempt to mediate between the conscious world of streets and street-names and the spatial unconscious

19 For more on the enduring influence of the counter-culture on MOORE see GRAY 2011.

20 Constantine was created by MOORE in 1985 as a secondary character in *Swamp Thing*. Constantine acted as guide and advisor on the supernatural for the titular character. Subsequently, Constantine has been written and drawn by a large number of creators for his own comic title, *Hellblazer*. Ironically, considering how well established and long-running the title is, MOORE has never written for *Hellblazer*.

of the Métro and the lost passages of the arcades (BUSE et al., 54). Unsurprisingly, BENJAMIN spoke of "threshold magic", symbolic portals located

> [a]t the entrance to the skating rink, to the pub, to the tennis court, to resort locations: *penates*. The hen that lays the golden praline-eggs, the machine that stamps our names on nameplates, slot machines, fortunetelling devices [...] these guard the threshold (BENJAMIN 1999, I1a, 4).

Once again, objects of the everyday can be turned on their head and re-enchanted. Such resacralisation reclaims quotidian space. The occult cartography of *Snakes and Ladders* follows MACHEN and takes the reader to the transitional, transitory space and time of "Moontown."

"Everyone knows this is Moontown, that has ever dreamed or wondered" (see Figure 2.3). This hallucinatory concatenation of image and text embodies and performs the spiralling motion of *Snakes and Ladders*. Although the two pages appear separate, they are linked by the iconic solidarity of the whirling shape of the spiral. The text, once again white on black, is interlaced between the jumble of images. The implied spiral of the left-hand page is rendered explicit in the mirrored right-hand page with CAMPBELL's inspired and uncanny looking superimposition of Edward Burne-Jones' ghostly angels from his painting *The Golden Stairs* onto the dual helix of DNA. Once again, CAMPBELL makes use of pro-graphic elements – images of film-stars, painter Richard Dadd, a frame from Meliès *Voyage dans la lune* (1902), even Vlad the Impaler – combine in a densely wrought pattern that calls on the reader to participate in the construction of meaning by making their own connections.

The constellation of seemingly incommensurable elements shatters any sense of stable temporality on the page, but at the same time the patterning gestures toward some kind of structure. "We have all reached the point of mixing up times" suggests Bruno LATOUR, adding that the spiral is a more apt model of time than the line:

> We do have a future and a past, but the future takes the form of a circle expanding in all directions, and the past is not surpassed but revisited, repeated, surrounded, protected, recombined, reinterpreted and reshuffled. [...] In such a framework, our actions are recognized at last as polytemporal (LATOUR 1993, 75).

Snakes and Ladders performs this mixing of times by its collaging techniques, which are themselves reflective of MOORE's allusive text. At the bottom of the helix, the right hand corner of the right hand page, is a panel from Winsor McKay's *Little Nemo in Slumberland* (1905–1927). The proto-Surrealist distensions of space and time in Nemo's adventures are always revealed to have been dreams, a return to reality with a bump. But Nemo's dreams retain their purchase in the nostalgic collective dreamscapes of popular culture.

NOSTALGIC SLUMBERLAND

This recalling of older forms, prior representations, historical figures, all points to an ambivalence regarding historicity. Nostalgia as a category has often been used as an insulting epithet, especially on the left. BONNETT argues that British psychogeography is marked as a site of conflict between two unresolved strands within political radicalism: "the use of the past to critique industrial modernity and the suppression of nostalgia." Nostalgia is a problematic category, expressive of an anxiety over the role of the past in radical politics. In proposing a political history of nostalgia, BONNETT presents an account "that shows how nostalgia emerged as a chronic problem and site of provocation within modernity" (BONNETT 2009, 47). In identifying this, BONNETT considers landscape as the "central stage for the proof and spectacle of transformation" (BONNETT 2009, 49).

External commentators have criticized London psychogeography for its perceived conservatism and voguish non-conformity. Journalist James HEARTFIELD attacked what he saw as the excessive romanticisation of the capital's urban landscape, which he termed "Londonostalgia." The London imagined by the psychogeographers is "a backward looking one" (HEARTFIELD 2004, n.p.). Phil BAKER writes that "anyone reading recent usages [of psychogeography] would discover that it is about Jack the Ripper, ley lines, why tower blocks are bad, Hawksmoor churches, the places we remember from earlier in our lives, landscape gardening, Stonehenge and the Kray twins" (BAKER 2003, 232).

In terms of critical theory, nostalgia "'is always someone else's problem" (BONNETT 2009, 53). An example of this is the difference between the postmodernist approach to history taken up by Linda Hutcheon and the idiosyncratic Marxist position adopted by Fredric Jameson: a spat marked by, as BONNETT puts it, "accusation and counter-accusation" that both view nostalgia as "a sign of failure and conservatism" (BONNETT 2009, 53).[21] Rather than attempt to read nostalgia as symptomatic of either the fashion plates of postmodernism or the revolutionary dreams of late Marxism, BONNETT proposes that the politics of loss are inherent in the politics of (post)modernity.

Nostalgia possesses both spatial and temporal qualities, manifested as a longing for a different, better time and place. Svetlana BOYM traces nostalgia to its roots as a part of a medical discourse in the seventeenth century and its definition as a curable disease whose first victims were displaced people: students, domestic servants, soldiers (BOYM 2007, 7–8). As BOYM points out, the "alluring object of nostalgia is notoriously elusive [...] an affective yearning for a community with a collective memory, a longing for continuity in a fragmented world" (BOYM 2007, 10). Nostalgia is tightly imbricated with the experiences of modernity and contemporaneity, especially changing conceptions of time and space. Rapid transformations of industrialisation and urbanisation (and, we might add, de-industrialisation and gentrification) run parallel with collective yearnings for cohesion. BOYM distinguishes between two types of nostalgia – restorative and reflective. The former seeks to "conquer and stabilize time" (BOYM 2001, 50) and "signifies a return to the original stasis, to the prelapsarian moment" (BOYM 2001,

21 BONNETT's work is itself part of a developing wave of nostalgia studies. For orientation to this field, BOYM's work (2001; 2007) is indispensable.

15). The latter is more flexible and "cherishes shattered fragments of memory and temporalizes space"; reflective nostalgia is "ironic, inconclusive, and fragmentary" (BOYM 2007, 15). BOYM concedes that the two may overlap to a degree, "they can use the same triggers of memory and symbols, the same Proustian madeleine cookie, but tell different stories about it" (BOYM 2007, 15).[22]

The association of radicalism and modernization occured, according to BONNETT, around the end of the nineteenth century with the rise of scientific socialism and the marginalization of the past as a resource for contemporary critique (BONNETT 2009, 50–51). But if the result was a hostility to nostalgia that lingers as a radical orthodoxy, then BONNETT suggests that "the past became more attractive to the unorthodox" (BONNETT 2009, 51). The dilemmas and opportunities of a radical, reflective nostalgia attracted the avant-garde and later the counter-culture of the 1960s and 1970s.

This has resulted in nostalgia acting as an alien, repressed presence within the radical tradition, and this problematic is played out in contemporary London psychogeography. This has acted as a "creative space in which feelings of loss and redemption are explored and negotiated" (BONNETT 2009, 54). Nostalgia is both invoked and repudiated into two strands that BONNETT associates with Iain SINCLAIR on the one hand, and the "revolutionary psychogeography" of groups like the London Psychogeographical Association on the other.

This raises the question as where to place MOORE and his work within this schema. Intuitively, he appears to have more in common with SINCLAIR. But this may not be so straight forward. What different psychogeographical practices seem to share is occult mapping as a method of re-working instrumentalised landscapes to resist the lure of heritage. What differentiates them is the role of the occult. For SINCLAIR, what is required is a counter-conjuration to combat what he labelled the "occult logic of 'market forces'" that dictate new urban geographies (SINCLAIR 1991, 265) For the LPA, the occult was enlisted as a method of provocation, ironic or otherwise, to aestheticise politics. In one famous missive, collected by Stewart HOME in *Mind Invaders*, the LPA announce that; "We offer no attempt to 'justify' or 'rationalize' the role of magic in the development of our themes; it is sufficient that it renders them completely unacceptable" (cited in COVERLEY 2008, 130).

The whole point of MOORE's occult practices, however, is that they are out in the open, bypassing the occult equivalent of the magic circle. This performativity is intimately linked with the visual field. The visual, especially in the construction of space and spatial re-presentation, was an important category for both the Situationists, and their forebears the Surrealists, as well as BENJAMIN. For the Situationists, the spectacle congealed capitalism within an image. For BENJAMIN, the dialectical image offered possibilities of "[t]elescoping the past through the present" and reconnecting with history (BENJAMIN 1999, N7a, 3). The dialectical

22 Cf. BONNETT 2006, 28: "I draw a distinction between nostalgia that tends towards the particular and nostalgia that offers a more abstracted sensibility of loss."

image brings the past and present into contact, seeking to "retrieve an object, practice or figure from obscurity. Its rearrangement in a new constellation offers the possibility of its transfiguration" (BUSE et al. 2005, 31).

CONCLUSION

As a psychogeographical text, Snakes and Ladders presents, or rather re-presents and re-evaluates, an occult cartography of London. A characteristic trait of British psychogeography is the tension between the immanent and the transcendent, the everyday and the numinous. The immanent planes of material place and material page, embodied in conventional and functional cartographic representation such as A–Z street guides, are reconstituted via graphic and textual collage. This is a historiography that reconstellates historical fragments to show how the past is always in a process of becoming. This recombinant process is, nevertheless, a melancholy act, as BENJAMIN recognized, seeking to conjoin existent and diverse fragments into an illusory whole. GROENSTEEN's idea of braiding remains suggestive when looking at the repetition of content throughout a comic which produces associations across the multiframe of a text. These associations in *Snakes and Ladders* point toward doublings and repetitions.

Psychogeography is defined by its doublings and repetitions, repeated in the axes of spatio-temporal movements, of the horizontal motion of walking coinciding with the vertical temporal descent into history. Rather than taking this at face value, it is perhaps better grasped as an act of re-constitution via montage to generate a sense of profane illumination. MOORE disinters the past of the area, pointing none too subtly to the repetition of exhumation in the history of Red Lion Square. That which is buried and repressed always returns. Concurring with BONNETT, BAKER concedes that psychogeography's attraction is precisely its backward looking-ness, of taking "refuge in interstitial zones of private meaning and esoteric knowledge […] an attempt to maintain a psychic investment in the street" (BAKER 2003, 332).

It is a nostalgia that inheres within the collaged constellative form itself, radical and reflective, revivifying faded fragments, attempting to reawaken a dormant sense of historicity associated with place. This affective and sensual experience is something that SINCLAIR proselytizes throughout his work, best articulated in his experience of walking around the London Orbital:

> The point of the day, the walk, is to lift that grey lid, the miasma of depression that hangs over the city and its inhabitants. To wait for the moment when the sun breaks through […] The remission. The pay-off that makes urban life worth living (SINCLAIR 2003, 29).

MACHEN's journey through the phantasmagorical Baghdad brings him into Syon and his own remission in light drenched modern day London (see Figure 2.4). The luminous glow takes over the hyperframe, effectively displacing the darkness. This is, nevertheless, distinctly ambivalent. The incandescent page, gesturing toward, but not quite achieving, total blankness, indicates erasure, and perhaps obliteration. It is this transcendent glow that the reader takes away (or is taken away by), the sublime backwash of history. And as MACHEN returns to sun dappled Holborn, we find ourselves, circumspectly, and conveniently, back at Red Lion Square, Conway Hall, 1999, April 10th.

REFERENCES

AUSTIN, J. L. (1962): *How to Do Things With Words*. Clarendon.

BAKER, P. (2003): Secret City: Psychogeography and the End of London. J. KERR and A. GIBSON (Eds.): *London: From Punk to Blair.* London, 323–333.

BENJAMIN, W. (1999): *The Arcades Project.* H. EILAND and K. MCLAUGHLIN (Trans.). Cambridge, MA.

BENJAMIN, W. (1999a): *Walter Benjamin: Selected Writings.* M. A. JENNINGS, H. EILAND and G. SMITH (Trans.). Cambridge, MA.

BONNETT, A. (1988): Situationism, Geography, and Poststructuralism. *Environment and Planning D: Society and Space* 7, 131–146.

BONNETT, A. (2006): The Nostalgias of Situationist Subversion. *Theory, Culture & Society* 23 (5), 23–48.

BONNETT, A. (2009): The Dilemmas of Radical Nostalgia in British Psychogeography, *Theory, Culture & Society* 26 (1), 45–70.

BUSE, P., K. HIRSCHKOP, S. MCKRAKEN and B. TAITHE (2005): *Benjamin's Arcades: An unGuided Tour.* Manchester.

BOYM, S. (2001): *The Future of Nostalgia.* New York.

BOYM, S. (2007): Nostalgia and Its Discontents. *The Hedgehog Review* 9 (2), 7–18.

COHEN, M. (1993): *Walter Benjamin and the Paris of the Surrealist Revolution.* Berkeley.

COVERLEY, M. (2006): *Psychogeography.* Harpenden.

COVERLEY, M. (2008): *Occult London.* Harpenden.

DARK, J. (1999): Occult Mysteries in Conway Hall. http://www.libra-aries-books.co.uk/pixie-inc/dark/ananke.html (accessed March 10, 2011).

DAVIES, O. (2012): *Magic: A Very Short Introduction.* Oxford.

DEBORD, G. (1955): Introduction to a Critique of Urban Geography. K. KNABB (Ed./ Trans.) (1981) *Situationist International Anthology*. Berkeley, California, 5–7.

DEBORD, G. (1967, 1983). *The Society of The Spectacle.* Detroit.

DEBORD, G. and WOLMAN, GJ (1956): A User's Guide to Détournement. KNABB, K. (Ed./Trans.) (2006): *Situationist International Anthology*. Revised edition. Berkeley, California, 14–21.

DI LIDDO, A. (2009): Transcending Comics: Crossing the Boundaries of the Medium. J. HEER and K. WORCESTER (Eds.): *A Comics Studies Reader*. Jackson, MI.

DITTMER, J. (2010): Comic Book Visualities: a Methodological Manifesto on Geography, Montage and Narration. *Transactions of the Institute of British Geographers* (35): 222–236.

DURING, S. (2002): *Modern Enchantments: The Cultural Power of Secular Magic.* Cambridge, Massachusetts.

FOSTER, H. (Ed.) (1988) *Vision and Visuality.* Seattle.

GRAY, M. (2011): Alan Moore's Underground: The Formation of a Dissident Cultural Practice. *Studies in Comics* 2 (1), 21–37.

GROENSTEEN, T. (1999, 2009): *The System of Comics.* B. BEATY and N. NGUYEN (Trans.). Jackson, MS.

HARVEY, D. (1973): *Social Justice and the City.* London.

HEARTFIELD, J. (2004): Londonostalgia. http://www.heartfield.org/Londonostalgia.htm (accessed July 17, 2012).
HOME, S. (Ed.) (1997): *Mind Invaders: A Reader in Psychic Warfare, Cultural Sabotage and Semiotic Terrorism.* London.
INWOOD, S. (2000): *A History of London.* London.
KNABB, K. (Ed./Trans.) (2006) *Situationist International Anthology*. Revised edition. Berkeley, California.
LATOUR, B. ([1991]1993) *We Have Never Been Modern,* C. PORTER (Trans.). Cambridge, MA.
MACHEN, A. (1894): 'The Great God Pan' in *Late Victorian Gothic,* R. LUCKHURST (2005): 183–233.
MACHEN, A. (1923): Extract from *Things Near and Far.* C. PALMER (Ed.): *The Collected Works of Arthur Machen*. London, 323.
MACHEN, A. (1924): *The London Adventure, or the Art of Wandering.* London.
MACKILLOP, I. (1986): *British Ethical Societies*. Cambridge.
MCDONOUGH, T. (2009): *The Situationists and the City*. London.
MILLIDGE, G. (2011): *Alan Moore: Storyteller*. Lewes.
MOORE, A. (1996): *Voice of the Fire.* London.
MOORE, A. (2006): Unearthing. I. SINCLAIR (Ed.) (2006): *London: City of Disappearances.* London, 328–373.
MOORE, A., S. BISSETTE, J. MCMANUS, J. TOTLEBEN, R. VEITCH and S. WOCH ([1984–1989] 2009): *Saga of the Swamp Thing* (six volumes), New York.
MOORE, A. and E. CAMPBELL (1999): *From Hell: Being a Melodrama in Sixteen Parts.* London.
MOORE, A. and E. CAMPBELL (1999a): The Birth Caul. E. CAMPBELL (Ed.): *A Disease of Language.* London. No pagination.
MOORE, A. and E. CAMPBELL (2001): Snakes and Ladders. E. CAMPBELL (Ed.): *A Disease of Language.* London. No pagination.
MOORE, A. and M. JENKINS (2010): *Unearthing*. London. (Music CD)
MOORE, A. and K. O'NEILL (2010) *The League of Extraordinary Gentlemen: Century 1910.* Marietta, Georgia.
MOORE, A. and T. PERKINS (1996a): *The Birth Caul.* Newcastle-upon-Tyne. (Music CD)
MOORE, A. and T. PERKINS (2000): *The Highbury Working, A Beat Séance.* London. (Music CD)
MOORE, A. and T. PERKINS (2002): *Angel Passage.* London. (Music CD)
MOORE, A. and T. PERKINS (2003): *Snakes and Ladders.* London. (Music CD)
MOORE, A. and B. SIENKIEWICZ (1990): *Big Numbers.* Northampton. (Two issues)
SCHWARTZ, J. M. and J. R. RYAN (2003): *Picturing Place: Photography and the Geographical Imagination.*
SINCLAIR, I. (1975, 1979, 1995). *Lud Heat and Suicide Bridge.* London.
SINCLAIR, I. (1991): *Downriver (Or, The Vessels of Wrath); A Narrative in Twelve Tales.* London.
SINCLAIR, I. (1997): *Lights Out for the Territory: 9 Excursions in the Secret History of London.* London.
SINCLAIR, I. (2003): *London Orbital.* London.
SINCLAIR, I. (Ed.) (2006): *London: City of Disappearances.* London.
VALENTINE, M. (1995): *Arthur Machen*. Bridgend.
VENEZIA, T. (2011): Edited City: London Passagenwerk. The Comics Grid, April 4 2011. http://www.comicsgrid.com/2011/04/edited-city-london-passagenwerk/ (accessed July 17, 2012).
WEINREB, B., C. HIBBERT and J. KEAY (2008): *The London Encyclopedia.* Third edition. London.
WILLIAMS, R. (1977): *Marxism and Literature.* Oxford.
WOODCOCK, P. (2000): *This Enchanted Isle: The Neo-Romantic Vision from William Blake to the New Visionaries.* Glastonbury.

Chapter 3 | *Shaun Huston*

LIVE/WORK: PORTLAND, OREGON AS A PLACE FOR COMICS CREATION

People are nonplussed when they find out you draw comics here.
"Well, yeah. Who doesn't?"
Everyone does.

David Hahn (writer-artist, *Private Beach, The Allnighter*)
In *Comic Book City, Portland, Oregon USA* (2012)

INTRODUCTION

Over the past quarter of a century – give or take – Portland, Oregon (USA) has become a 'place to be' for comics creators. Unlike other such places, notably New York City, which is a center for the global publishing industry and home base for both Marvel and DC Comics, Portland's attractiveness for comics writers and artists is not grounded in a well-established industrial economy. That lack raises interesting questions as to why and how Portland is not only host to a critical mass of comics creators, but one that continues to grow.

In my documentary film, *Comic Book City, Portland, Oregon USA* (2012), I construct a representation of the city as a place for comics creation based on interviews primarily with writers and artists, but also publishers, editors, and editorial staff, who have chosen to live and work in Portland. In this chapter, I highlight key aspects of this representation and elaborate on the broader contexts for understanding the significance of the city for people in this field. While comics creation and comics creators can be located in relation to broad categories such as the creative class (FLORIDA 2002) or "neo-bohemia" (LLOYD 2006), in both this chapter and in my documentary I emphasize the particularity of the relationships between people and place in Portland. Indeed, the city's attractiveness for comics writers and artists is indicative of the ways in which Portland is deployed both as an exemplar of development models derived from broader discourses on cities and creativity (see, for example, FLORIDA 2005, PECK 2005), and as a cautionary case (RUSSELL 2011, SILPAYAMANANT 2012).

I begin with a brief discussion of how the documentary was conceived and made. In the section that follows, I look at the broader historical and geographic contexts for comics creation as an industry and for the (re)location of creators to Portland in particular. From this general overview, I move to a more specific examination of what my research suggests as to why writers and artists have been relocating to Portland and why they choose to stay. In the conclusion, I consider the permanence or durability of the city's comics scene.

THE FILM

Comic Book City began as a way for me to intertwine my cultural geographic interests in place with my interests in film study. I began making short films in 2005, primarily as a way to experiment with concepts of the frame, the shot, and, especially, editing. These are aspects of filmmaking that DEAR (2011) would refer to as, "the creativity *of* place" in art, or, "the ways in which space itself is an artifact in the creative practice or output, as when a dancer moves an arm through an arc or a photographer crops an image to create a representational space" (9). In other words, I undertook these films as formal experiments related to film theory, and not with the intent of exploring and representing cultural relations of people in place. However, when I turned my attention to possible feature-length works, I sought to address those relationships as well.

I was doing this thinking in 2007, and likely as a result of looking for a suitable topic or theme, I started to notice the number of creative ties between comics I was reading and Portland, notably that publishers and many of the individual creators were based in the city. I also noticed that these ties crossed lines between kinds of comics: web and print, monthly serials and graphic novels, corporate and creator-owned.

My initial investigations suggested that comics in Portland would be a promising topic for documentary work. Press reports such as *Creativity and Comics Thrive in Portland*, by Chris ULLRICH at *Comic Book Resources* (2007), suggested that the city had an established reputation as a place where people go to make comics (cf. WATERHOUSE and JAMAN 2007). I also learned of the existence of Periscope Studio, an independent collective of writers and artists in downtown Portland with a current membership of twenty-seven. The presence of an independent work space like Periscope and the noted press reports, combined with what I had previously observed regarding publishers, which include Dark Horse Comics, Oni Press, and Top Shelf Productions, confirmed that there was a particular question about 'creativity *in* place' (DEAR 2011) to be explored between comics creators and the city of Portland.

Michael DEAR defines 'creativity in place' as, "the role that a particular location, or time-space conjunction, has in facilitating the creative process" (9). DEAR cites as examples, turn of the century Vienna, Silicon Valley and Hollywood, but his description also adheres to Portland in thinking about comics as a creative vocation and as an industry. *Comic Book City* is a representation of how comics creators understand the city as a place for making (or from which to make) comics.

As I began to shape a narrative, one focusing on those directly involved in the creative process, I decided that it would be important for the film to include commentary on Portland in historical and geographic perspective, and for that I interviewed Carl ABBOTT and Charles HEYING, both faculty in urban studies and planning at Portland State University (PSU), and Shanna Eller, at the time, a PhD student in urban studies at PSU. ABBOTT has written a number of histories of Portland (for example ABBOTT 2001; 2011), while HEYING and Eller had been working on a project examining the city's 'artisan economy' (HEYING 2010).

I conducted the on-camera interviews between April and September 2008. In addition to recording interviews during this period, I also shot footage of places and events, including the Stumptown Comics Fest, 24-Hour Comic Day at Cosmic Monkey Comics, and Periscope Studio. I also collaborated with photographers Erin Marr and Patrick Huston to create a bank of still photographs of the city and collected work samples from the featured artists to be incorporated into the film.

The finished documentary is structured around eight writer/artist interviews. I selected interviews on the basis of a subject's camera presence and to show a range of experiences and the diversity of the community, particularly in terms of sex and gender and kinds of work, that is, writing, drawing, writing and drawing, working for hire, working independently, working in long form, working in serials, web publishing and print publishing. Artist and writer sections are connected by contextual segments featuring the academics from Portland State and also broader examinations of recurring themes, including the influence of Dark Horse Comics, living without a car, and Portland's 'whiteness'. Video of locations and events, still photographs, and art samples are used to signify different aspects of the creative process, particularly as practiced in place, and qualities or attractions of the city as discussed in the featured interviews.

HISTORICAL-GEOGRAPHIC CONTEXT

Comics creators can be fruitfully seen as participating in what Charles HEYING calls Portland's 'artisan economy' (2010). In a general sense, HEYING defines this economy as one in which the worker engages in productive activity with, "hands, head, and heart" (40). Salient characteristics of this activity include the production of objects made: by hand, for appreciation as well as consumption, and in egalitarian fashion, without regard to the lines between high and low culture (41–44). Artisan work is defined by flexible schedules and becomes both, "an identity and a vocation" (45). The artisan economy is organized on a small scale, commonly through 'live/work' spaces like the archetypal artist's loft, and through social networks and community interactions that promote cooperative or shared learning and recognition from within the artisanal social world (46–49). HEYING locates the artisan economy in opposition to the Fordist model of mass production, mass consumption, and large scale corporate organization, and sees the emergence of artisan economies as a grassroots and creator-led adaptation to post-Fordist conditions, notably decentralization and flexible production and specialization.

As noted by HEYING (2010), the culture of comics creators in Portland reflects and reproduces key attributes of the artisan economy, particularly in the way that artists, writers, publishers, editors, readers, and shop owners constitute a mutually supportive community. It is also the case that as work and as a creative act, making comics takes place in small spaces, in home studios, at coffee shops, at individual desks and tables in workplaces like Periscope or Dark Horse, and entails work by hand, work that is meant to, "intentionally reveal the touch of the maker, and honor the inherent qualities of the material being shaped" (HEYING 2010, 41). Notably, even in popular monthly comics like those published by Marvel and DC, the style of the artist and voice of the writer are valued attributes (and commodities) to readers and publishers (cf. WOLK 2007). As an expressive form, comics are also on the edge between the fine arts and popular culture. Doing comics well requires some of the same skills as that of a painter or writer of literature, but results in a product that is egalitarian or accessible in the way HEYING defines artisanal work (cf. NORCLIFFE and RENDANCE 2003).

The importance of handwork is one reason that comics creators are better located among HEYING's artisans than in Richard FLORIDA's creative class, which primarily produces ideas or concepts that are then turned into objects, or put into use, by others (2002). Comics creators certainly do conceptual and intellectual work when designing characters or making stories, but, in the end, a comic is an object, whether in paper or on a screen, one crafted by the creators, and not by someone else. It is the rare circumstance when a writer or artist would be employed solely to create an idea or concept and not to also make the comic.

In many cases, comics creators work in a way similar to those whom FLORIDA calls creative professionals, that is, workers tasked with solving particular problems, but without expectation for the origination of generally useful knowledge. That latter job belongs to those in the super-creative core of the creative class (2002). A writer or artist effectively works as a 'creative professional' in circumstances where they write or do art on a title owned by a publisher, or other rights holder, and where the characters are not original to the individual creator, or creators, but executing their job nonetheless requires devising particular ways of drawing a character (or characters) and their world, and placing them in new stories.

While there are writers and artists who only work on other people's creations, and this would be common for specialists in fields like inking, coloring, and lettering, many comics creators also do original work alongside work-for-hire. As an economic class, then, comics creators are in-between the lines of stratification identified by FLORIDA, but more significantly, whether doing original work or working with others' creations, making comics entails crafting an object, and does not stop with, or primarily involve, the production of knowledge or information meant to be transferred to other workers.

FLORIDA's creative class is also notable for being part of the social elite and economic driving force in the cities where its members live and work (cf. PECK 2005). However, making comics is not, in and of itself, a pathway to even a comfortably middle class life, let alone a source of income that results in significant economic power. Most comics creators derive financial support and income from a variety of sources. Many do other kinds of writing or artwork, including writing straight prose or criticism, doing illustration, and making storyboards for film and television. Some writers and artists teach part-time at colleges and universities. Sales of prints and original art is another supplemental income stream. Some have non-comics related jobs, or spouses or partners working in stable careers (cf. NORCLIFFE and RENDANCE 2003; EDROSO 2011).

While Portland features well in cultural terms from a creative class perspective – FLORIDA ranks the city at the top of a list of American cities that are 'child-friendly', and it also scores high on his 'Creativity Index' and 'Gay Index' (2002) – the city's economic infrastructure does not demonstrate key qualities identified with a strong attractiveness to individuals in this class. In particular, as observed by Charles HEYING, the city lacks the "extensive venture capital system" that FLORIDA cites as critical for growing the creative economy (2010, 295–296; cf. ABBOTT 2001).

ABBOTT characterizes Portland as a "provincial city", one where change and development happens more slowly than in comparable places such as Seattle and Las Vegas. He argues that the surrounding landscape of forests and mountains, and the frequent thick cloud cover, has cultivated "a sense of limits" (2001, 10, 12–13). This sense of limits is politically expressed in the city's urban growth boundary and regional government, Metro, which is charged with measuring, planning, and managing the urban region's expansion into the countryside (ABBOTT 2001).

Along the same lines, HEYING identifies a "less is more" ethic as part of the "complex moral economy" of Portland's artisans (2010, 49–50). The limited economic prospects to be had from comics suggests not only that choosing to work in this field almost requires an embrace of 'less is more', but also that there are distinct advantages to living and working in a city that is, essentially, "a sort of poor person's Bay Area" (ABBOTT 2001, 93). For example, at the time of my interviews in 2008, all but one of the eight writers and artists featured in *Comic Book City* owned homes, and the attractiveness of being able to afford the purchase of a house while still making comics is cited repeatedly by creators as a reason to be in Portland.

The need, and willingness, to live with limits, with less as more, is another characteristic that distinguishes comics creators, and Portland's other artisans, from FLORIDA's creative class, or, even, Richard LLOYD's "neo-bohemians", both of whom are defined by strong commitments to opportunity seeking and geographic mobility or to seeing locations like Portland, or Chicago's Wicker Park in the case of LLOYD's case study, as potential launching grounds for bigger and better places and things. In contrast to fields like the traditional visual arts, theater, or film, there is no place where comics creators go to 'make it big', at least not as comics creators. To the extent that Portland 'works' for individuals with aspirations or established careers in comics, there are seemingly few reasons to leave once roots have been planted.

In their study of North American comic book production, NORCLIFFE and RENDANCE find that the mainstream comics industry has undergone a transformation in the organization of production, dating from the 1980s, wherein centralized 'bullpen'-styled production has been replaced by a more dispersed model of employing freelancers who may be located in different parts of the world from each other and from the companies that employ them. These individuals are not only not kept in-house, but may never meet face-to-face or directly with editors and publishers. Major publishers, notably Marvel and DC, have become, essentially, "marketing operations" (2003, 252), selling products that are, more often than not, actually produced by strings of individuals who likely also work on titles from other companies and on their own independent and creator-owned projects. NORCLIFFE and RENDANCE note that these changes have simultaneously enabled the deconcentration of comics production from major centers like New York, and also the clustering of producers in new locations, including smaller towns and cities. The accretion of writers and artists in Portland clearly demonstrates these trends. My conversations with creators in *Comic Book City* offer a picture of why Portland has become a prominent place in the contemporary geography of comics creation.

MOVING TO PORTLAND

The individual creator featured in *Comic Book City* with the longest term of residence in Portland, writer-artist Paul Guinan (*Heartbreakers, Boilerplate*), moved to the city from Chicago, Illinois in 1992. For Guinan, and other longer term residents like Kevin Moore, (writer and artist, *In Contempt* and *Wanderlost*, moved from Buffalo, New York in 1995), a central influence on their relocation to Portland was the presence of Dark Horse Comics. This influence can be direct, in Guinan's case his spouse had been hired as an editor by the company, or indirect. In Kevin Moore's example, the location of Dark Horse in Milwaukie, a suburban town adjacent to Portland's southeast side, was confirmation of his developing impression that Portland was home to an emergent comics community.

As LLOYD notes in his Wicker Park study (2006), specific artistic and intellectual scenes come to be represented to wider publics by selected examples of success. For Portland and comics, Dark Horse is the original example. The publisher is repeatedly noted and made notable in media reports for its longevity (founded in 1986), and, depending on the measurement, for being either the third or fourth largest U.S.-based publisher after Marvel and DC. Dark Horse is also noted for having pioneered a business model based on both creator-ownership and licensed properties. In terms of the local scene, many reports also note the shaping influence of Dark Horse both in terms of recruiting talent to the city and spinning it off (cf. ULLRICH 2007; TURNQUIST 2008; CRAWFORD 2009; ARNOTT 2010; MIRK 2011).

The basic elements of this narrative are recapitulated in my documentary, not just by Guinan and Moore, but also by Dark Horse owner and publisher Mike Richardson, and by Dark Horse editors Scott Allie and Diana Schutz, and former editor, Shawna Gore. This recapitulation affirms the company's significance not just from media reporting but also in terms of how those within the community of comics creators understand the history of the local scene.

One detail worth adding to this story is the fact that Richardson, unlike every other individual interviewed in *Comic Book City*, is a Portland-area 'native', and his attachment to the city is a particular reason for why Dark Horse is, and remains, in Milwaukie. He cites the city of Portland and its environs as assets for recruiting and retaining creative people. In a recorded interview segment that I did not include in the final film, he describes a "grand tour" for prospective employees and potential creative partners designed to showcase the city's urban amenities, as well as the ready access to mountains and coast. This strategy echoes FLORIDA's emphasis on "quality of place" for the creative class (2002, 231–234; cf. ABBOTT 2001).

From my interviews with writers and artists, Dark Horse now appears more in the background than in the foreground of reasons for setting up in Portland. Younger and more recent residents featured in the documentary, including Sarah Oleksyk (creator of *Ivy*, moved to Portland from Portland, Maine in 2001), Dylan Meconis (creator of *Bite Me!* and *Family Man*, moved to Portland from Seattle in 2005), and Graham Annable (creator of *Grickle,* originally from Sault-Ste. Marie, Ontario and moved to Portland from the San Francisco Bay Area in 2005), do not cite the company when discussing their reasons for moving to the city, nor do they note the publisher in their personal geographies of Portland's comics culture.

Rather than a particular entity or institution serving as a draw, more recent residents note the presence of other creators, both individually and en masse, and including those already considered to be friends and colleagues, as a primary reason for choosing Portland. Carl ABBOTT touches on this dynamic in the documentary when he remarks:

> It's kind of interesting how some cities become hot, exciting places that people begin to talk about, and the fact that people talk about them makes them interesting and then interesting people come and visit or decide to settle down, and then it becomes more interesting because there are more interesting people there.

I found this pattern reflected in the stories of many of the comics creators I spoke with for the documentary. More than one recalled becoming aware of Portland as a possible place to live after talking to other writers and artists at conventions – for example – and noting how many of their friends and colleagues were already in the city.

In the film, Dark Horse editor Scott Allie also makes note of the way in which Portland's comics culture has become in some sense detached from the publisher as the population of writers and artists has grown.

> I went into Guapo Comics, this comic book shop on Foster [Road in southeast Portland], for the party the night before Stumptown [Comics Fest; more on this below]. I walk into this party and the store is crammed full of people. It's shoulder-to-shoulder, and I barely recognize anybody. They're all doing zines, or maybe better production than that, and to them, they're the world. They know there's this Dark Horse thing out here, but, like, "Oh yeah, those are color comics, I really don't like color comics." And that's amazing. That's so great, that they have this really vibrant scene that's totally independent of us, that has nothing to do with us.

While Dark Horse may be at the base of Portland's development into a 'comics town', the continuation of that development appears to be more dependent on wider word-of-mouth and media reputation than on a single institution.

Sarah Oleksyk expresses the significance of being with other creators as providing "validation" for her choice to work in comics, something she did not find in other places, including her hometown of Portland, Maine and New York, where she attended art school.

> I grew up in Portland, Maine ... There's support for the arts out there, but it's not my kind of art ... And then I came to New York, and I went to art school, and I didn't hardly know anybody else who was doing comics there. They didn't have a comics program, and every teacher told me it was a waste of time. I was still doing it, but I was doing it despite everybody else, instead of because of everyone else. I just didn't get any validation for what I had chosen to do, and it wasn't until I came out here that I got tons of validation from others .. Even if I don't show my stuff to other people, knowing that other people are doing and taking it as seriously as I do is really motivating ...

This resonates with HEYING's observation about the significance of internal recognition within the artisan economy, but in way that reads as distinct from the dynamic in LLOYD's Wicker Park, where the neo-bohemian milieu seemed to breed competition as much, or more, than camaraderie. Oleksyk continues:

> When there's a show people will invite me to be in it, and when there's a party going on where other cartoonists are going to be, and I'm there, I get to converse with people and talk about the business, and this just shows me that this is something beyond a shared hobby. And for a profession that is incredibly solitary, I've never had so many people that I've been working with, or working alongside ... If ever I have a problem, I can just show my friends – "What am I doing!? Help me out!" – and get more perspective on it. And getting congratulations when people get whatever milestones they reach. I never felt like I had peers until I moved out here.

These sentiments are clearly echoed by others in the documentary, including, in the case of writers and artists, by Kevin Moore and Sara Ryan (writer, *Me and Edith Head*, *Orienteering*, moved to Portland from Ann Arbor, Michigan in 1997), and by Shawna Gore, in the case of editors and publishers.

The 'validation' of the choice to work in comics is also understood by the creators I spoke with as being embedded in a larger culture of acceptance or openness for being different or doing things outside of the cultural mainstream As Steve Lieber (artist, *Whiteout*, *Alabaster: Wolves*, moved from Ann Arbor in 1997), puts it:

> Portland, as I'm sure other folks have said, is maybe the geekiest city in America right now ... I grew up ... in Pittsburgh, and that was not a place for someone like me. I'd carry my sketchbook out and sit in diners and start drawing. I'd get threats for just being there doodling in my sketchbook. In an environment like that, a geek, someone who likes to cartoon, that's one of those nails that sticks up that gets hammered down. Here, everyone is sticking up in some weird configuration.

Similarly, Sara Ryan sees Portland this way: "There's just a lot of openness in general to ideas and ways of living ... I like that it's okay to be sort of wonky in Portland. There's definitely a lot of sort of wonky, wonky and geeky, people and that seems to be good". Extending this line of thought, Dylan Meconis describes the city as a place where:

> There's the stereotype of everybody in Portland does at least three things. You know, you have your day job, and then you have your side passion, and then you have all the things you volunteer for, or all the things you serve on the board of, that you go to on Saturday nights … I've met very few people who have just their one thing, and ride it to the end of the rails. And every time I think I've found somebody normal, I find out how horribly wrong I am.

Graham Annable finds that, "Every artist here just kind of follows their own beat, and it's sort of encouraged. The city's sort of built a reputation on that from what I can see … I think people are kind of allowed the space and time to find themselves up here." The quote from David Hahn that opens this chapter, in effect that no one is surprised to hear that you make comics, also emphasizes the extent to which people with non-mainstream vocations are accepted by others in Portland.

Statements such as these can be read as suggestive of 'quality of place' as defined by FLORIDA, particularly in regards to the desire of people in creative fields to be in locations where there is, "an openness to diversity of all kinds, and above all else the opportunity to validate their identities as creative people" (2002, 218; see also 13–14 re: "the experiential lifestyle"). However, in Portland, and with this particular group of 'creatives', desires for openness and validation are clearly related to a specific sense of outsiderness that comes from working in a culturally marginal form like comics, and not from a general wish to be affirmed as 'creative' or valued for one's creativity in an abstract way.

STAYING IN PORTLAND

In a sequence before the opening titles to *Comic Book City*, the writers and artists I feature share their answer to the question, "Would you ever consider leaving Portland?" In different ways, and to different degrees, each says, essentially, "no".

As suggested previously, there are probable structural economic reasons why comics creators, particularly those who work as creative professionals in the industry, may find a city like Portland, which is "good for entrepreneurship, less so for income" (ABBOTT 2001, 203), to be a suitable, and, notably, affordable, place to establish and make a career. However, the feeling of community referenced in the section just prior also appears to play a vital role in grounding the writers and artists who have established themselves in the city.

The sense of "validation" that comes from being in a place with a critical mass of comics creators is not simply a matter of seeing others do what you do, but also lies in the cultivation of a sense of community with those individuals, what HEYING (2010, 48–49) calls the "soft infrastructure and learning communities" of the artisan economy (cf. LLOYD 2006, NORCLIFFE and RENDANCE 2003; Carl ABBOTT also makes a similar point in the film). For virtually all of the creators I spoke with for *Comic Book City*, the Stumptown Comics Fest serves as a primary expression of that sense of community and is a key feature of the 'soft infrastructure' that supports making comics in the city through active forms of communication and exchange.

The first Stumptown Comics Fest was held in 2004, as an artist-led and volunteer-run festival, which it remains. The event features panels on art and industry, portfolio reviews for aspiring creators, signings, and vendor tables for publishers, individual writers and artists, and comics stores. David Hahn explains how the festival originated this way:

> There were people here, like "The Little Rascals".
>
> "Hey, let's put on a show!"
>
> "Yeah, a comic book show!"
>
> Well, okay, let's go rent this space, and do some advertising, and I'll make the fliers and you call artists up and see who wants to buy a table. And people make it happen, like the Stumptown Festival. That's, wow, people get up and they do things here.

This view is also reflected in Kevin Moore's explanation of what makes Stumptown different from other 'cons':

> It's not like San Diego [Comic-Con International], which is pretty much taken over by Hollywood, but it's not like A.P.E. [Alternative Press Expo], which is a great convention unto itself, but which is much more alternative press oriented, obviously in its name. But it's not like your kind of flea market, basement thing that goes on with a lot of comic cons. It's a real great blend of artists and publishers, most of them very independent, but some of them corporate, but we're getting together and we're sharing our ideas, and our perspectives, and our artistic sensibilities.

What both creators are touching on here is a spirit of community, of cooperation and sharing, as exemplified by Stumptown.

For many of the individuals I spoke with for the documentary, the festival represents this community ethic, the *learning* community (see HEYING 2010), that makes being in Portland meaningful, and not just a matter of convenience or economic rationality. Shawna Gore sees the event this way:

> The Stumptown Comics Festival is sort of the perfect Portland thing to me because it's a great showcase for all this local talent and local publishers, artists, crafty people, arts and crafts … It's not like we're [Dark Horse] – a great big corporate sponsor who just writes a check. We have people on our staff in editorial who make their own mini-comics and who have comics-related blogs and they set up their own tables at Stumptown. So, we have people who work for Dark Horse set up there in an entirely different capacity, and we have people set up behind the Dark Horse table.

Kevin Moore adds this understanding:

> I met a bunch of cartoonists who, ironically enough were coming from the east coast … And we've learned a lot from each other, whether it's new things to read, new developments going on in the industry, new techniques. We all do different things. We're not just underground cartoonists, or superhero cartoonists. We're everything. Most of my friends, we just have huge, diverse interests in comics and cartooning, and we come together with our different sensibilities, and I think have a really positive and enriching effect on each other. And I think you'll see that reflected in the Stumptown Comics Fest.

This sense of exchange, of variety and cooperation, is indicative of a community of creators who see each other as colleagues more than as competitors or as models for the 'right way' of 'making it'.

As an event run by creators, and without a conventional corporate structure, Stumptown, even though it brings in exhibitors and guests from other places, clearly benefits from the density of writers and artists in Portland. It is possible to stage the event relying solely on panelists, speakers, and vendors who are locally convenient. Not surprisingly, as expressed by Paul Guinan, the Fest is one of those 'only in Portland' things he would miss if he were to leave the city.

While Stumptown appears to be a common touchstone in the expression of community around comics in Portland, other signs of the city's 'soft infrastructure' for sharing knowledge, ideas, and opportunities noted in the film include: Periscope Studio, the variety of publishers, the city's wealth of comics shops, formal and informal events like 24-Hour Comic Day, and a general cooperative and collegial spirit amongst writers and artists that leads to hanging out and working in people's homes and at places like pubs and coffee shops (cf. HEYING 2010, NORCLIFFE and RENDANCE 2003).

Beyond comics and community specific reasons for staying in Portland, the individuals I interviewed for *Comic Book City* cite a variety of 'quality of place' reasons for not just moving to the city, but putting down roots. A primary example of these reasons is the ability to live and work without owning a car, which not only figures into how individuals understand the comparative cost of living in the city, but is also reflective of the way in which the Portland is broadly seen as supportive of different or alternative ways of living, particularly ones that are locally-driven, 'progressive', and 'sustainable' (cf. HEYING 2010).

CONCLUSIONS

Over the closing credits to *Comic Book City*, I play excerpts from my interview with Dylan Meconis wherein she reflects on the present moment and looks to the future.

> Somewhere like Seattle back in the '90s was a really huge comics scene, and now there are still a lot of people up there, but all of a sudden Portland is the new focus town. So I recognize that the scene is not going to be this crazy forever. People are going to move. Some other town, you know, Tucson! is going to become the next big independent comics town, but for right now it's really remarkable to have all of these people coming here at the same time. I think that Portland's current oddball attitude and the vibrancy of the scene are very connected. It's not permanent, but it's wonderful to be here while it is that way.

Read one way, these comments can be seen as part of what Richard LLOYD refers to as "the structural nostalgia", or the "sense of being always already over", in neo-bohemia (2006, 237). This is particularly true of Meconis's implication that there are those moving to Portland right now just to 'make the scene', and who will be gone as soon as there is a new scene to make.

On the other hand, her reflection suggests more of a slowing down in growth than a fading or decline. Furthermore, whereas the 'structural nostalgia' observed by LLOYD is rooted in a feeling that neo-bohemian places are ruined by too much attention, Meconis is celebrating the current 'craziness', not bemoaning it, while also expressing that there is more to than city than its current fashion. As I highlight in the opening to the film, this confidence in Portland as 'home' is shared by other creators.

HEYING speculates that Portland's artisan economy, particularly in the making of things, exists not in spite, but because, of the city's high unemployment since 2008. Even though the region has shown one of the weakest job markets in the U.S. during the current recession, people continue to move to the city (HEYING 2010), and local media continue to report more specifically on the movement of comics writers and artists to Portland (for example, LIPSKI 2012). Whether in the thoughts and feelings of current residents, or in stories and data about migration to the city, Portland persists as a place for comics creation, perhaps becoming another historical example of an urban region, like Hollywood or Silicon Valley, that is culturally identified with a particular form of 'creativity in place'.

REFERENCES

ABBOTT, C. (2001): *Greater Portland: Urban life and landscape in the Pacific Northwest.* Philadelphia.

ABBOTT, C. (2011): *Portland in Three Centuries: The place and the people.* Corvallis.

ARNOTT, L. (2010): The Top Five Comic Book Publishers: Biggest comics companies, May 22, 2010. http://suite101.com/article/the-top-five-comic-book-publishers-biggest-comics-compan ie-a239745 (accessed June 26, 2012).

CRAWFORD, W. (2009): The Rise of Comic Town, OregonBusiness, December 2009. http:// www.oregonbusiness.com/articles/74-december-2009/2634-the-rise-of-comic-tow (accessed June 26, 2012).

DEAR, M. (2011): Creativity and place. M. DEAR, J. KETCHUM, S. LURIA and D. RICHARDSON (Eds.): *Geohumanities: Art, history, text at the edge of place.* New York, 8–18.

EDROSO, R. (2011): The Comics Issue: If cartoons are so big, why don't they pay? *The Village Voice*, April 6, 2011. http://www.villagevoice.com/2011-04-06/news/the-comics-issue-if-cartoons -are-so-big-hy-don-t-they-pay/ (accessed June 26, 2012).

FLORIDA, R. (2002): *The Rise of The Creative Class: And how it's transforming work, leisure, community and everyday life.* New York.

HEYING, C. (2010): *Brew to Bikes: Portland's artisan economy.* Portland.

LIPSKI, C. (2012): Portland Welcomes New Resident Chris Roberson. *Examiner.com*, January20, 2012. http://www.examiner.com/article/portland-welcomes-new-resident-chris-roberson (accessed June 26, 2012).

LLOYD, R. (2006): *Neo-bohemia: Art and commerce in the postindustrial city.* New York.

MIRK, S. (2011): A Brief History of How Portland Became the Best Comics City in America. *The Portland Mercury*, July 28, 2011. http://www.portlandmercury.com/portland/a-brief-history-of-how-portland-became-the-best-city-in-america-for-comics/Content?oid=4384408 (acessed June 27, 2012).

NORCLIFFE, G. and O. RENDANCE (2003): New Geographies of Comic Book Production: The New Artisan, Distancing, and the Periodic Social Economy. *Economic Geography* 79 (3), 241–263.

PECK, J. (2005): Struggling With the Creative Class. *International Journal of Urban and Regional Research* 29 (4), 740–770.

RUSSELL, J. (2011): Pittsburgh Vs Portland. Burgh Disapora, July 28, 2011. http://burghdiaspora.blogspot.com/2011/07/pittsburgh-versus-portland.html (accessed June 27, 2012).

SILPAYAMANANT, J. (2012): Portland is Where Young People Go to Retire. *Mae Mai*, January 29, 2012. http://silpayamanant.wordpress.com/2012/01/29/portland-is-where-young-people-go-to-retire-the-curse-of-the-creative-class/ (accessed June 27, 2012).

TURNQUIST, K. (2008): Portland Comics Publishers go Hollywood. *The Oregonian*, July 22, 2008. http://www.oregonlive.com/entertainment/index.ssf/2008/07/kristis_new_comx.html (acessed June 27, 2012).

ULLRICH, C. (2007): Creativity and Comics Thrive in Portland. *Comic Book Resources*, July 23, 2007. http://www.comicbookresources.com/?page=article&old=1&id=11255 (accessed June 26, 2012).

WATERHOUSE, B. and C. JAMAN (2007): Drawn Together. *Willamette Week*, September 26, 2007. http://www.wweek.com/portland/article-7647-drawn_together.html (accessed June 27, 2012).

WOLK, D. (2007): *Reading Comics: How graphic novels work and what they mean.* Boston.

PART 2 | BODIES POLITIC

Chapter 4 | *Catriona MacLeod*

FROM WANDERING WOMEN TO FIXED FEMALES: RELATIONS OF GENDERED MOVEMENT THROUGH POST-COLONIAL SPACE IN *LETTRES D'OUTREMER* AND *LE BAR DU VIEUX FRANÇAIS*

The recent commercial and critical success of Marguerite Abouet and Clement Oubrerie's *Aya de Yopougon* series[1] has, since the publication of its first volume in 2005, brought the female figure of non-European origin to the forefront of the French-language comic strip, or *bande dessinée*. Despite the visibility of Aya and her Ivoirian friends in bookshops and Francophone comics festivals since 2005, however, such representations of women of African origin remain rare in the *bande dessinée* – a medium which has, throughout its history, overwhelmingly favoured depictions of white male characters. In his article "The Representation of Ethnic Minority Women in Comic Books", Mark McKinney introduces his thesis by noting the importance of the *bande dessinée* as a cultural form in France and Belgium, attuned to the socio-political developments affecting, and of interest to, its relatively young demographic. Where women of ethnic minority or immigrant origin are depicted, he then contends, these representations "provide significant insights into the ways in which gender and ethnicity have been constructed in popular culture" (McKinney 2000, 85). McKinney goes on to note, however, that women of non-European origin are not only underrepresented in the *bande dessinée* but that in strips in which they do appear, they are "regularly relegated to the margins" (87). Accepting as valid McKinney's assertion of the *bande dessinée*'s socio-cultural value, where such female figures do exceptionally appear as leading characters in the Francophone comic strip, then, close analysis may provide particularly interesting results.

This chapter aims to compare the representation of two female characters of non-European origin in *bandes dessinées* by white male artists.[2] Both Souana,

1 This six-volume narrative follows the life of the eponymous Aya, a young woman from the Ivory Coast. The series has sold more than two hundred and fifty thousand copies and has been translated into twelve languages (Gallimard n.d.). Its first tome was awarded the prestigious "Prix du premier album" (Award for Best First Album) at the 2006 Angoulême festival.

2 Although *Aya de Yopougon* boasts a female creator of African origin (writer Marguerite Abouet is originally from the Ivory Coast), this constitutes another example of this series' exceptional nature. The great majority of *bande dessinée* creators are male – in 2009, Thierry Groensteen estimated that approximately ninety percent of *bande dessinée* creators actively working in the industry were men (Lemaire 2009) – and amongst male creators, most are from white European backgrounds.

who appears in WARNAUTS and RAIVES' *Lettres d'Outremer*[3] (1996) and Leila, a figure from LAPIÈRE and STASSEN's diptych *Le Bar du vieux Français*[4] (1992; 1993) occupy prominent roles opposite male characters in their respective tales.

Lettres d'Outremer and *Le Bar du vieux Français* present via the medium of the *bande dessinée* two differing incarnations of the post-colonial travel narrative. The former text charts the story of Jean, a Parisian journalist, who is sent to French overseas province Guadeloupe to write an article on the island for "Paris-Max" magazine. There he meets and engages in a relationship with Antillean beauty Souana, who acts as his unofficial tour guide. STASSEN and LAPIÈRE's two-part *bande dessinée*, in contrast, follows the parallel journeys of two runaways – a young African boy, Celestin, and a teenage girl of Maghreb origin, Leila – as they head north and south respectively, in search of better lives. The comparison of texts with such apparently different relations to movement immediately highlights a complex issue. Although the terms "travel" and "migration" have traditionally been considered as referents of distinct practices – "travel" as a luxury largely restricted to the Western middle and upper classes and "migration" as a necessary activity undertaken for economic or political reasons, associated with peoples living outside of the First World – considerable attention has been turned in recent theorisations of the postcolonial travel narrative to the decolonisation and democratisation of the concepts of travel, migration and exile (FORSDICK 2005, 162). In "Travelling Theory, Exiled Theorists", Charles FORSDICK notes the growing acceptance that "migrancy and travelling are complementary as opposed to competing experiences" (FORSDICK 2001, 7). The following parallel analysis of *Lettres d'Outremer* and *Le Bar du vieux Français* is undertaken with this recent recognition of the potential similarity between, and the decolonisation of, patterns of movement in mind. As the previous qualification attests, stories of postcolonial travel and migration have, in recent years, received considerable academic attention from scholars in the fields of French and Francophone studies as movement between France and Belgium and their former colonies continues to prove a complex but unavoidable element of post-imperial reality in the French-speaking world. Little work has, however, been undertaken yet concerning manifestations of such narratives in the *bande dessinée*. This chapter integrates these two previously disparate fields of study by considering the trope of travel in *Lettres d'Outremer* and *Le Bar du vieux Français*, with a particular focus on its relation to the principal female character of each tale. In addition to studying the interrelation of gender and ethnicity in these examples of popular culture, then, this comparative analysis will also assess how these interwoven elements are connected to the contrasting notions of migration and stasis, and, finally, to the less tangible but nonetheless recurring notion of unattainable desire throughout the post-colonial journey.

3 "Letters from Overseas"
4 "The Old Frenchman's Bar"

LETTRES D'OUTREMER

The Belgian creative pair WARNAUTS and RAIVES have produced more than thirty *bande dessinée* albums since their partnership began in the mid-1980s, with many of these narratives taking the reader to foreign and far-flung locations. *Lettres d'Outremer* chooses as its principal setting a place caught between exoticism and Francophone familiarity: the French-Antillean island of Guadeloupe.

Principal character Jean's visit to the island has layered motives. Ostensibly, as noted above, he is sent to experience the island before writing an article that will be read by metropolitan French readers. Having just lost his wife to a sudden and fatal brain tumour, however, Jean's editor later admits that he was chosen for the overseas assignment in order to allow him some respite and a change of scenery following his recent bereavement. Jean's personal desire to go to Guadeloupe, the reader progressively understands, in turn extends beyond both of these motivations – having found evidence of an affair between his late wife and a Guadeloupian man in letters uncovered after her death (the eponymous "Lettres d'Outremer"), he wishes to find and confront this unknown Antillean lover in person.

Despite this complex web of reasons for Jean's presence in Guadeloupe, *Lettres d'Outremer* nonetheless repeatedly evokes the professional assignment of its principal character, as the narrative is punctuated by interludes introducing the reader to aspects of the island's culture and geography both visually, in silent frames which depict stalls at the local market or long, panoramic views of beach or forest, and textually, via, for example, explanations of local dishes written beneath frames in "handwritten" narrative script. Jean's position as a traveller to Guadeloupe charged with the task of investigating its exotic features takes a fortunate turn when he meets Souana early in the narrative. She soon becomes both his guide and a further object of Antillean curiosity to be explored.

The first encounter between Jean and Souana happens by chance at a party. Souana's representation is almost immediately visually sexualised as, after their short, introductory conversation, a sequence begins in which Jean watches her dancing with another man. In a technique repeated sporadically throughout *Lettres d'Outremer* (and in other of WARNAUTS and RAIVES' works) four frames in increasing close-up, akin to photographic snapshots, focus progressively on Souana's swaying form as her partner is incrementally cropped from each image. The reader occupies Jean's viewpoint as Souana looks out of the frame while she dances, her coy smile inviting a sexualised gaze in return. The narration below this sequence underlines Jean's burgeoning attraction as it describes her movements: "Son corps tourne dans la nuit caraïbe. La robe, serrée à la taille, s'évase sur les cuisses. Les morceaux de tissu fendus laissent voir ses jambes longues et fines, des éclairs de chair dorée. Il pense qu'elle doit avoir un corps superbe"[5] (WARNAUTS and RAIVES 1996, 18).

Jean and Souana's second encounter, although also by chance, commences Souana's role as the former's island tour guide. The pair pass each other on the street in the town of Pointe-à-Pitre and, after inviting herself to join him in his intention to head to the beach for a swim, Souana takes Jean on his first Guadeloupian bus journey, advising him when to pay as Jean recounts his first experience of an Antillean drink, *chodo*, at the recent party. Once at the beach, despite Jean's earlier consternation that she is equipped with neither swimwear nor a towel, Souana hurries to the water in her vest, calling for her new acquaintance to join her. A frame of Jean, swimming underwater on his front, shows he has obliged, while the suggestive pairing of this image with a frame directly beneath it of Souana floating on her back foregrounds their burgeoning sexual relationship (27).[6]

Overleaf, a pivotal scene commences the development of Souana's symbolic role in the narrative. Following four frames showing the pair chatting in the water, during which Souana asks Jean candid questions about his late wife, a fifth ocean image suddenly shows Souana alone, centred within a frame, surrounded by the sea (Figure 4.1). The late-afternoon sun is directly behind her and this arrangement of light casts shade over her body, presenting her usually-defined features as an indistinct golden blur. Reflected in the rippling water, Souana's hazy outline melds with the natural scene around her, suggesting her body to be one with the island scenery. Her words, "Je suis curieuse, hein?"[7], placed in a speech-bubble directly above her head and ostensibly pronounced in acknowledgement of her probing questions to Jean about his wife, are transformed amongst this striking visual scene into an enunciation of Souana's status as an object, rather than bearer, of curiosity. Both text and image combine in this frame to present Jean's new companion as an exotic "natural feature" of the island to be experienced and displayed for the French reader.

From this iconic image, Souana becomes progressively linked to an idyllic, exoticised view of Guadeloupe, as experienced through the eyes of incoming Parisian traveller, Jean. In between assisting him in his quest to find his wife's mysterious lover, Souana introduces Jean to her favourite beauty spots around the island, with WARNAUTS and RAIVES dedicating entire pages of frames to detailed views of Antillean natural scenery. A double page of such images mid-way through the narrative underlines Souana's increasing connection to Guadeloupian

5 "Her body turns in the Caribbean night. Her dress, fitted at the waist, flares at the thighs. Splits in the fabric show off her long slender legs in flashes of golden flesh. He thinks she must have a superb body."

6 For an analysis of scenes relating to Souana and Jean's intimate relationship and a study of Souana as a modern incarnation of the symbolic "Black Venus" see article "Post-Colonial Persistence: The Presence of the "Black Venus" in the works of Warnauts and Raives" (MACLEOD 2011).

7 "I'm curious, eh?"

nature. The first page in this sequence, following on from Souana's promise to show Jean an exceptional location, is entirely silent. Six frames show rich brown and green-toned views of dense tropical forest. In the final image, Souana is seen (once again from Jean's viewpoint) disappearing into the forest. Her head is turned to look out of the frame but long hanging leaves cover her eyes, this restricted gaze emphasising her position as the object of Jean's look, much like the natural scene around her. Jean, standing behind Souana in this photographic perspective, is not shown in these images.

The opposite page shows the pair at their destination: a waterfall hidden in the forest. This island idyll is showcased by WARNAUTS and RAIVES in a large introductory panoramic frame which occupies half the page (Figure 4.2), while the narration above poetically describes the scene as "un cocon vegetal transpercé par un soleil au zenith"[8] (67). In this sizeable image again only Souana is visible, perched on a rock looking out towards the waterfall wearing only her underwear. In the three frames that make up the strip below, Jean appears in the middle image, emerging from the pool at the foot of the falls to declare the location as "ce que j'ai vu de plus beau depuis que je suis arrivé"[9] (ibid.). In the frames on either side, Souana remains silent, watching Jean from her position on the rocks, making up the final element of this most beautiful of natural scenes.

Souana's symbolic connection to Guadeloupe extends beyond the visual "naturalisation" of her physical form, however, as an idealised vision of island existence is sporadically also presented through her dialogue. Her first words in the narrative are a favourable comparison of Guadeloupe to Paris, from where she is noted to have just, happily, returned. Later in the diegesis her commentary progresses to paint, at times, a nostalgic picture of Guadeloupian life. During another sequence in which Souana has taken Jean to see a place of natural beauty, for example – a fishing village by the sea, dutifully depicted by WARNAUTS and RAIVES in a series of richly-coloured frames –, she explains to Jean: "Les gens d'ici vivent suivant le rythme d'un autre temps, tu sais...un veritable osmose avec la mer..."[10] (59). The vision of Guadeloupe that Souana verbalises is simple and harmonious, evoking for Jean – and the reader – an idealised image of another era.

In her work *Stories of Women: Gender and Narrative in the Postcolonial Nation*, Elleke BOEHMER (2005, 4) notes the intriguing reappearance across time and across nations of women as the bearers of national culture. This tendency is particularly visible, she states, during the twentieth-century process of decolonisation and the corresponding creation or re-establishment of independent nation states in which – although historically existing as male-constructed spaces, brought into being by male leaders and activists – "women are more often than not cast as symbols or totems, as bearers of tradition" (22). The character of Souana, in contrast, appears both visually and verbally throughout *Lettres d'Outremer* as the

8 "a cocoon of plants pierced, at its height, by the sun."
9 "the most beautiful thing I've seen since I arrived"
10 "The locals here live by the rhythm of another time, you know...in real harmony with the sea".

bearer of a neo-colonial nostalgia, rather than new-found post-colonial identity, presenting to the French visitor Jean, and the largely Francophone, European *bande dessinée* readership, a simplified and idealised view of Guadeloupe. The vision of an island utopia is inscribed upon her body while her speech describes an uncomplicated, campestral culture. In "Ethnic Minority Women in Comic Books", MCKINNEY (2000, 91) attests to the persistence of neo-colonial representations of women of African origin in a selection of *bandes dessinées* which illustrate, he claims, the "sexist dimension of France's Eurocentric colonial heritage." In *Lettres d'Outremer*, the neo-colonial implications of Souana's depiction extend beyond the reproduction of the colonial relationship between white male protagonist/reader and black exoticised female to replay this imbalanced connection between the European spectator and the island of Guadeloupe.

Souana's words describing Guadeloupe do not constitute the only discourse concerning the island in the *bande dessinée*, however, nor is Souana the only Antillean character with whom Jean has contact while on the island. Near the opening of the narrative, Jean makes the acquaintance of Antoine, a Guadeloupian journalist, with whom he strikes up a friendship. Antoine in many ways becomes the symbolic counterpart to Souana's idealised presentation of Guadeloupe throughout the narrative. During the first conversation of Jean and Antoine in the story, Antoine evokes both the physical problems of the island – the environment, the roads, detritus in the streets – and the postcolonial problems provoked by Guadeloupe's status as both an Antillean island and an official region of France, describing this position as "schizophrenic" (WARNAUTS and RAIVES 1996, 38).This conversation is directly followed, in contrast, by a sudden change of scenery and a series of four frames which show Souana's hands at each stage of preparation of a traditional Antillean "ti-punch" cocktail, this reinforcing both the travel-guide motif of the album and Souana's place as bearer of exotic culture. Later in the story, Antoine appears again as he accompanies Jean to an incident involving the French riot police, during which Jean is struck in the head by an overzealous white policeman who asks him, as he tries to record the event, whose "side" he is on (76). Immediately following this sequence, while Antoine drives Jean to a doctor for medical attention, the heavens open and *Lettres d'Outremer* presents its first and only scenes of grey sky and tropical rain. As heavy drops fall, Antoine again textually evokes the difficulties of modern Guadeloupe – the problems with healthcare, unemployment, and the patronising attitude of metropolitan France to its overseas territories – in frames which show neither him nor Jean but simply feature speech bubbles filled with dialogue amid sodden urban scenes (77). Significantly, however, in the proceeding frames which depict Souana inspecting Jean's head wound, the rain has stopped and sun shines over Guadeloupe once again.

The clear contrast between the roles played by Antoine and Souana in *Lettres d'Outremer* emphasises two important aspects of the latter's symbolic function. Firstly, the evocation via Antoine of the rather dystopian socio-economic reality of Guadeloupe underlines the impossibility – or at least the exaggeration – of the idealised island view that Souana represents. In *Nowhere is Perfect: French and*

Francophone Utopias/Dystopias, Jean WEST-SOOBY describes the concept of "utopia" as dialectical: on the one hand it presents an ideal model of land or life, whilst on the other it becomes a pointed reminder of the imperfections the utopia has itself been designed to abolish (WEST-SOOBY 2008, 2). Antoine's presence in *Lettres d'Outremer* thus serves to underline the negative corollary of the utopian dialectic, rendering Souana's representation of and connection to the Antillean island idyll (sadly) fleeting and ultimately unreal.

It is also via the juxtaposition of Antoine and Souana's contrasting roles that the importance of the latter's position as woman – exoticised for the male character and reader alike – to her links with the nature of Guadeloupe and the corresponding nostalgic view of its culture becomes apparent.[11] Although, as noted, some of Souana's dialogue is devoted to the idealisation of her island home, it is mostly via visual representations of her body amongst nature that her symbolic role is made clear. Antoine, on the other hand, provides his opposing interpretation of Guadeloupe predominantly textually, in the only long passages of dialogue in the narrative. As the male corollary to Souana's bodily connection to their island home, Antoine's role is clearly verbally, rather than physically, focused and centres around a consistently politicised verbal discourse.[12] The last frame of Souana in the story places a final emphasis on her woman-ness as integral to her symbolic role. Following a series of silent images which show her at the beach with Jean, Souana appears alone in the centre of a frame which recalls her earlier shadowy, sea-bordered representation (Figure 4.3). She stands at the water's edge in a traditional blue smock which obscures her now well-known curves but with her delicately crossed legs and the poise of her outstretched arms underlining her femininity. Her position at a certain distance from the reader blurs, to an extent, her features; however, a speech bubble directly above her head clearly pronounces

---

11 Clear links can be drawn between Souana's role in *Lettres d'Outremer* and the feminising of colonised territories in colonial discourse. In *Feminism and the Colonial Body*, Kadiatu KANNEH (commenting on a notion enunciated by Franz Fanon) discusses the representation of Algerian women over the course of French imperial presence in their country "as scopic signifiers of colonised Algeria itself – as oppressed, inscrutable, and dispossessed" (KANNEH 1995, 346). In Souana's case, her beauty, warmth and sexualised exoticism are extended to the physical and cultural notion of Guadeloupe that is presented to the reader.

12 The hybrid nature of the comic strip, whose narratives are a product of the fusion of text and image but whose formal specificity often structurally separates the two (text is more often than not encased in a delineated speech bubble, or may be placed outside the frame) renders it a medium particularly effective in the divisive visual/textual focus between genders noted in *Lettres d'Outremer*. This split is notable in other of WARNAUTS and RAIVES works and, on occasion, is manipulated to divide characters across racial as well as gender lines. In "Congo blanc", for example, the first short graphic narrative of collection *Equatoriales* (1994), the bodily exoticism of the principal black female character is underlined by the lack of a corresponding textual presence, a feature significantly not mirrored in the depiction of her white Belgian lover, or, to an extent, the latter's European wife.

the first and last words of Antillean Creole uttered by a character in the text.[13] This last image, which so clearly connects the character to Guadeloupian nature and tradition, presents not "Souana" in this symbolic role, but "woman." It is the form of an Antillean female, not the recognisable outline of Souana, that the reader sees before them, this female figure's statuesque pose rendering her a timeless, immovable symbol of the island around her.

As noted, Souana is not seen again in the narrative. Following this beach sequence, Jean is shown to be back in Paris, having completed his magazine article. He speaks to Antoine on the phone and welcomes a visitor from Guadeloupe to his home, but the only indication of contact with Souana comes via a letter she has written to him, delivered by his guest, whose contents are never revealed (but which constitute his own *lettre d'outremer* from an Antillean lover, indicating the completion of his internal, personal journey). Whilst the physical or verbal presence of other characters may travel to Jean's side in Paris, Souana it seems, remains immobile, fixed to the island she represents. Her depiction throughout the narrative, from the indication that she has just returned from metropolitan France to the Antilles at the start of the story, progressively anchors her to the land that she increasingly symbolises. Jean's decision in the final frames of the album to fly back to Guadeloupe (we assume to rejoin Souana, although this is never stated) emphasise once more his position as traveller, contrasting his free mobility with Souana's now-established stasis.

LE BAR DU VIEUX FRANÇAIS

Whilst *Lettres d'Outremer* presents a travel motif divided between genders, with its principal female figure static in relation to the transatlantically-mobile male protagonist, Jean-Philippe STASSEN and Denis LAPIÈRE's two-part *Le Bar du vieux Français* recounts the parallel journeys of a boy, Celestin, and teenage girl, Leila, who travel respectively north and south through Africa. Both are shown in motion during the tale and it is the narrator, the eponymous and never-named *vieux Français*[14], who remains stationary, telling the story of the young pair to customers in his bar.

The story opens with scenes depicting Celestin in his home village, a rural settlement somewhere in Africa. In order to protect his sister, Kudi, from the tribal violence which has overtaken his once peaceful home after the death of their parents, Celestin leaves the village with his sister to walk towards the nearest town. Kudi dies during the difficult desert journey and Celestin continues on throughout the narrative, stopping for sometimes long periods to work in various places, heading ever north. Leila's introduction to the story places her in a very different

---

13 These words are translated into French below the image as "Vois comme la vie peut être douce...Moi j'arrive après le dernier orage pour t'apporter l'embellie" ("See how life can be sweet...I arrive after the last storm to bring you the sun").

14 "Vieux Français" translates simply as "Old Frenchman."

home environment: a family of North-African origin living in a European city.[15] Longing for freedom away from the controlling presences of her father and brother, Leila sneaks out of her house one night and does not return. She travels through Spain to Morocco and it is during a journey even further south through the desert that she crosses paths with Celestin at the bar owned by the *vieux Français*.

Whilst Souana is established early in *Lettres d'Outremer* as a bearer of idealised Guadeloupian culture, this association only strengthening as the narrative progresses, Leila's inaugural scenes in *Le Bar du vieux Français* depict her as a young woman who rejects her traditional cultural origins. The first frame in which she is present graphically illustrates this notion as Leila is shown vomiting in the bathroom of her family home while the traditional sacrifice of a sheep for the festival of Eid is carried out downstairs (LAPIÈRE and STASSEN 1999, 19). Refusing to leave the bathroom at her mother's request, it is only when her father commands her to show respect for the day of celebration that she reluctantly obeys, glowering as she passes through the house at the separate gatherings of men and women (20). Over the next few pages, the reader sees Leila rebel further, this time turning her back on the strict, gendered constraints of her upbringing. She discusses wearing a mini-skirt to school with her friend Anne, is caught smoking in the school toilets by her brother and tries to leave the house alone to meet her friends – an attempt thwarted by her father who indicates that neither Leila nor her sisters have permission to go out unaccompanied (32).

The "runaway" is a distinguishable trope amongst representations of women of North-African origin in the *bande dessinée*, with MCKINNEY citing in his aforementioned article examples of female characters who, like Leila, abandon their family homes in order to escape patriarchal domination by male relatives. In his work MCKINNEY links this trope to "the dominant assimilationist perspective, inherited from colonial times, that second-generation Maghrebi women are cultural lynchpins who do or should abandon their foreign cultural heritage in order to help align their group with mainstream cultural norms" (MCKINNEY 2000, 88). Far from embodying an idealised view of traditions and culture whilst anchored to her land of origin (i.e., Souana), then, in the events that lead up to her night-time "escape" from her parents' adopted home Leila is cast as a figure longing to discard her cultural heritage – presented over these few pages in a distinctly negative light – to embrace the freedom symbolised by her white European friend Anne.[16]

15 Artist STASSEN has noted Leila's home town to be that of Liège, Belgium, although this is not made clear in the narrative (MCKINNEY 2000, 99).

16 The un-nuanced comparison of Leila's "oppression" and Anne's "freedom" at the beginning of *Le Bar du vieux Français* renders this *bande dessinée*'s critique of patriarchal structures in the North African family problematic. MCKINNEY notes that the presence of rigid or oppressive depictions of Maghrebi gender relations in certain *bandes dessinées* are used to maintain a structure of "positional superiority" between representations of European and North African societies (MCKINNEY 2000, 93). MCKINNEY later relates this to *Le Bar du vieux Français* specifically when he notes that the presence of "a critique of traditionalism, especially patriar-

Despite the clear links at the narrative's beginning between Leila's cultural origins and the male domination she wishes to flee, the restrictive portrayal of elements connected to the character's heritage does not extend to the geographical origin of these cultural practices. Leila leaves home with Anne's help and heads simply "south", the first stop on her journey taking her to Barcelona. After a short stay and a chance meeting with a young Frenchman who, enamoured by her beauty, offers to be her chauffeur, Leila leaves this city to head towards her desired destination: Morocco.

Following her arrival in Tangier alone in the car she has stolen from the young Frenchman,[17] Leila's depiction changes. In the earliest scenes representing this character, she is often shown in confined spaces pressed into narrow frames, her physical environment emphasising her experience of familial oppression. The first image which shows Leila's face (following its hidden view as she is unwell during Eid) depicts her petite form curled up on the bathroom floor, squashed between the door and the toilet, as she snaps at her mother in the hall to go away. The first frames of Leila's journey into Morocco, however, as she drives off the Mediterranean ferry and into the countryside, contrast vividly with this established mode of representation: for the first time, Leila is shown amongst green fields and wide open spaces. These introductory frames of Moroccan scenery are followed by an entirely silent page of images (one of only two in the narrative to be presented devoid of dialogue or narration) which shows Leila sitting by the sea under a sunny sky, smiling as the wind blows through her hair. A large panoramic frame introduces this sequence, casting Leila as a tiny figure on the rocks watching the waves as the reader gazes at this scene from a distance behind the featured figure (Figure 4.4). Comparable to the image in *Lettres d'Outremer* showing Souana perched before the waterfall, Leila here becomes one with the scene around her, her small stature disappearing into the landscape. Remaining largely unsexualised in this frame in contrast with the corresponding view of Souana depicted in her underwear, Leila's blue jeans and white top serve to meld her more completely with the natural view before her, matching the deep blue water and white ocean spray.

Leila's connection to the land of Morocco is, at first, contrasted with her isolation from its apparently gendered culture. She is hassled soon after she disembarks the ferry by an over-eager petrol attendant who notes his surprise that she is

---

chy, associated with Islam, in a Moroccan-Belgium family is not extended to the surrounding Euro-Belgian society, which by implication is uncritically equated with women's liberation" (ibid.).

17 Leila will go on to steal another car from a German traveller later in the story. This repeated theft of vehicles from white voyagers – although somewhat "justified" in both cases by the behaviour of each European figure (the French man puts pressure on Leila to sleep with him whilst the German woman strikes her) – suggests a symbolic seizing of the act of travelling by the figure of African origin from the white tourist. It is comparable to a narrative trope in J.R Essomba's *Le Paradis du Nord* (1996) which sees the Cameroonian protagonist Jojo fund his plans to travel north to Europe by stealing from wealthy tourists who stay in the luxury hotel in which he works (LOINGSIGH 2003, 159).

not scared to be driving alone, and her aforementioned reverie in front of the sea is interrupted when she becomes aware that she is being watched by an unknown boy. The clearest expression of Leila's isolation from traditional Moroccan life comes directly after this sequence, however, when she drives to the village of her extended family. Expecting a warm welcome from her aunt, uncle and grandfather, Leila is, instead, greeted with stares of shock and immediate consternation that her father allowed her to leave home alone (55). Standing before her uncle on the first page of this sequence, the *vieux Français*' narration recounts that "ils se dévisageaient l'un l'autre comme s'ils venaient d'une autre planète"[18] (ibid.). Fleeing this second oppressive family, Leila's car is hit by a lorry at the very moment she begins to wonder whether she should abandon her journey and return to her father. The shock of the accident swiftly wipes this notion from her mind, however, and Leila's first thoughts as she lies beside the wreckage of the stolen car announce her intention never to return to her Francophone country. Abandoning the wreckage, and thus leaving the last connection to her old, restricted life behind her, the reader understands in the final frame of this sequence that Leila is reborn following this traumatic experience. An idyllic, panoramic scene shows her walk away through a grassy valley against a sky lit by shades of a red and yellow sunset as the narration of the *vieux Français* notes: "Il fallait se rendre à l'évidence comme si le choc lui avait remis la tête à l'endroit. Leila était tombée amoureuse de ce pays"[19] (LAPIERE and STASSEN 1999, 61). Despite – and, paradoxically, due to – her final separation from its traditionally patriarchal culture, this image links Leila to the country of her origins, contrastingly rooting her freedom in its landscape.

This symbolic connection of Leila to the physical manifestation of Morocco is less marked, and certainly less sexualised than its equivalent in *Lettres d'Outremer.* Whilst Souana is repeatedly placed in the midst of striking natural scenes, often wearing little clothing, the linking of Leila to her geographical "homeland" is more understated. It is, however, reinforced in certain subtle ways. It is notable that, for example, despite the ostensibly comparable nature of Leila and Celestin's journeys in *Le Bar du vieux Français*, place names are used more frequently in relation to Leila's voyage. Celestin's nationality is never revealed (although in a scene which shows him hand over an identity card to suspicious policemen towards the end of the narrative, the flag visible on the card appears to

18 This translates as: "They stared at each other as if they were from different planets." The character of North African origin returning to their (or their parent's) country after an extended period in Europe – and the potential difficulty of this experience - is a trope explored in other *bandes dessinées*. In CHRISTIN and GOETZINGER's *La Voyageuse de Petite Ceinture* (1985) – an earlier narrative featuring a young Maghrebi runaway– the principal character's brother recounts his visit to extended family in Algeria and his struggle to relate to their lifestyle – one, as he puts it, with "no rock music, no McDonalds, no mates, nothing" (CHRISTIN and GOETZINGER 1985, 15).

19 "She had to admit it, as if the shock had brought her back to her senses. Leila had fallen in love with this country!"

be that of Burkino Faso) and none of the destinations in which he stops alone on his journey are named. Leila's travels are much more clearly documented. The town in which she first stays is indicated briefly to be Barcelona, and, following this short interlude, it is Morocco and its northern city Tangier that are the most clearly and repeatedly demarcated.

Additionally, whilst Celestin's journey takes him continually north, never returning even briefly south of his last location, Leila's travels circle Morocco. Following her revelatory car crash, Tangier is her first destination. Arriving in the town on foot, she is immediately captivated by "les odeurs et la fraicheur des coins sombres"[20] (LAPIÈRE and STASSEN 1999, 69). Spending the night there, the next morning the narration recounts that "dès l'aube, elle se jeta dans la ville, avide de découvrir d'autres quartiers, d'autres regards"[21] (70). A chance encounter with a German tourist in need of a driver to take her south gives Leila the chance to earn some much-needed money and so she leaves Morocco on a journey that will take her through unidentified desert territory, ending up at the isolated *bar du vieux Français* alongside, for the first time, Celestin. After a brief stay and another incidence of car theft, however, Leila returns north to Tangier with Celestin. Later in the narrative, after a further short trip south, Leila again finds her way back to Tangier – the Moroccan city continually calling Leila back as Celestin, in contrast, forges consistently north.

Leila's connection to the land of Morocco – and the strangely paradoxical freedom from her cultural origins that this link symbolises – does not last throughout *Le Bar du vieux Français*, however. Although Leila's representation within the tale is predicated from the start on her physical and mental journey away from home, this migratory freedom is incrementally reduced towards the end of the narrative as, correspondingly, views of the character before beautiful landscapes and intriguing city spaces disappear. After spending some time in Tangier with Celestin, the pair befriends a Moroccan boy, Abdelaziz, who procures a receptionist job for Leila at the "Hotel de France." Whilst Celestin leaves the city to continue his journey north, Leila's relationship to movement now appears to invert, as she progresses from being a voyager herself to becoming a stationary welcoming presence to incoming travellers, presumably – as indicated by the name of her employing establishment – from Francophone Europe. Even this inverted connection to movement disappears, however, as soon after, Leila agrees to marry Abdelaziz. She is no longer shown working and her married life is depicted in a series of frames which illustrate her sitting alone or listlessly chatting with female friends in the home she shares with her new husband (Figure 4.5). Correspondingly, following the sequence depicting the pair's wedding (discussed more fully below), Leila is never again shown outdoors and the names of her adored locations – Morocco, Tangier – are no longer uttered. The static and phys-

20 "The smells and the cool patches of shade."

21 "As soon as dawn broke, she ran out into the town, eager to discover other areas and other sights."

ically restricted nature of this new existence is finally clearly stated when Leila receives a visit from an acquaintance of Abdelaziz. After discussing the former's frequent absence due to his extensive business travels to Paris and New York, the friend announces "Ah Leila!...tu mérites mieux que cette espèce de vie cloitrée à laquelle tu t'obstines"[22] (117), an unambiguous verbal expression of her already visually-enunciated enclosed, motionless existence. Significantly, on the page following this poignant declaration, as the story winds to its close, Leila's narration takes over from that of the now-deceased *vieux Français*. Where before this old man assumed a stationary role between Leila and Celestin's parallel journeys, as the story ends Leila takes on this static function as Celestin travels further away.

It is clear that both the progressive halting of Leila's journey and the corresponding disassociation of the character with the exterior beauty of Morocco in *Le Bar du vieux Français* correlates with her gradual integration back into a patriarchally-focused society. It is her rejection of her father's authority at the start of the tale which initially motivates her journey, and it is following her similar experience with and flight from her strict Moroccan family that she is reborn "in love with" the landscape of her origins. After meeting Abdelaziz, a young man from a rich, established Moroccan family, however, and following the departure of Celestin – an orphan who has also fled from an (albeit very different) oppressive male regime – Leila's journey begins to stall and scenes of outdoors exploration are replaced with frames showing Leila in interior spaces. It is, as noted, Abdelaziz who procures Leila's job at the hotel and it is during a scene which shows her working behind the reception desk that Abdelaziz hurries in, begging her to marry him and announcing his intention to leave immediately for Europe to discuss matters with her father (112).

During the sequence of frames showing the pair's wedding, it must be noted that two rather humorous images do show Leila again as a young woman rejecting the traditional gendered culture of her origins, as she is depicted tipping drops of fake blood onto bed sheets while swigging champagne directly from the bottle (113).[23] These scenes are preceded by a long frame representing the wedding celebration, however, which foregrounds a group of men smoking and chatting, whilst Leila's mother and other female guests watch from the hall. This image recalls the first sequence in which Leila is present in the narrative, during which she scowls at the separate groups of men and women assembled for Eid. The re-

---

22 "Ah Leila! You deserve better than this sort of cloistered life you're clinging to".

23 Leila here is trying to circumvent the traditional wedding practice which sees the bed sheets of the newlyweds inspected by family members for drops of blood thought to prove the bride's chaste sexual past. Although this is noted by Kathleen Coyne KELLY still to be a feature of some North African weddings, particularly in rural areas, she also indicates that the depiction of this practice in Western popular culture is often used as a means of implicitly mocking or criticising the "misogynist" practices of Islamic cultures (KELLY 2000, 128) This tallies with MCKINNEY's concerns noted above regarding the relation of "positional superiority" identifiable in some *bandes dessinées* critical of North African patriarchal family structures.

production of a similar scene during her wedding clearly conveys her (albeit somewhat rebellious) re-entry into this world.

The final frame of the narrative cements Leila's integration into – and corresponding stasis within – the patriarchal system she previously rejected as she is shown, sitting motionless, holding a new-born baby. Her function as a woman in male-dominated society complete after giving birth, her journey is now at an end. She is physically and symbolically enclosed within traditional patriarchal culture. Her freedom – represented in her connection to, and love for, the land of Morocco – proves to be just like the idealised image of Guadeloupe presented by Souana: a fleeting desire, unable to be ultimately realised in the face of the dominant, dystopian reality.

CONCLUSION

Lettres d'Outremer and *Le Bar du vieux Français* are two very different examples of the *bande dessinée* post-colonial travel narrative. Both, however, manipulate representations of their principal female characters within and through space to a similar end. In Elleke BOEHMER'S aforementioned text, she cites the following view from German writer Klaus Theweleit: "Woman is an infinite, untrodden territory of desire which at every stage of historical deterritorialisation, men in search of material for utopias have inundated with their desires" (Theweleit in BOEHMER 2005, 22). Souana and Leila, in their respective texts depicting aspects of a post-colonial world after the deterritorialising processes of the late twentieth century, are imbued with expressions of *unattainable* desire. WARNAUTS and RAIVES render the beautiful Souana an embodiment of an idealised, simplified Guadeloupe. The contrasting of this figure with that of Antoine, whose dialogue exposes the reality of Guadeloupian life in its complex post-imperial existence, infuses Souana's representation with a nostalgic desire for an island paradise which cannot fully exist. The linking of Leila to the beautiful landscape of Morocco, established after she rejects the possibility of taking on the expected feminine role of "bearer of national culture", correspondingly diminishes as she is drawn via a rapidly unfolding narrative back into the patriarchal traditions of her country of origins, in a problematic neo-colonial critique of gender relations in modern Moroccan families. Her connection to, and mobility through, the North African landscape becomes representative of a desired "freedom" for women of Maghrebi origin which, for LAPIÈRE and STASSEN, simply cannot endure in the current post-colonial context. In both cases, the linking and restricting of the female figures to specific spaces is contrasted with the transnationally-mobile and unconstrained existences of prominent male figures whose freedom in the post-imperial world appears assured.

The comparative analysis presented here contributes to a developing focus of study concerning the links between gender and mobility within the post-colonial travel narrative. It also adds an intriguing element to this debate which has, previously, remained largely unconsidered: how the formal specificity of the *bande dessinée* may contribute to the configuration of such a link between gender and movement in depictions of journeys through post-colonial space. In both examples of the *bandes dessinées* analysed, the interaction of the textual and the visual is key to the progressive restricting of woman. In *Lettres d'Outremer*, it is the contrast between the pictorial connection of Souana to an idealised vision of Guadeloupe and the textual evocations between Antoine and Jean of the territory's problems which highlights the former's role as the static, eternal symbol of an ultimately unattainable island paradise. In *Le Bar du vieux Français*, it is the combination of the visual juxtaposition of cramped, interior frames and panoramic exterior scenes and the contrast between textual affirmations of Leila's liberty or restriction, which underlines the unsustainability of the character's "freedom" in the current post-colonial world.

In *Travel in Twentieth-Century French and Francophone Cultures* FORSDICK considers the relatively recent shift in French studies whereby "popular cultural forms have steadily replaced canonical literary texts as objects of study" (FORSDICK 2005, 158), further noting that "There is a need to recognise, however, that reading the *récit de voyage*[24] seems to involve more complex critical practices than those implied by the study of, for instance, the *bande dessinée* or *polar*"[25] (ibid). The analysis presented in this chapter suggests that the dual nature of the sequential art *récit de voyage* may in fact be a particularly effective means of depicting the complex challenges of the post-colonial journey, and one which merits more attention.

REFERENCES

BOEHMER, E. (2005): *Stories of Women: Gender and Narrative in the Postcolonial Nation*. Manchester.

CHRISTIN, P., and A. GOETZINGER (1985): *La Voyageuse de Petite Ceinture*. Paris.

FORSDICK, C. (2001): Travelling Theory, Exiled Theorists. C. FORSDICK (Ed.): *In Travel and Exile: Postcolonial Perspectives*. Liverpool, 1–13.

FORSDICK, C. (2005): *Travel in Twentieth-Century French and Francophone Cultures: The Persistence of Diversity*. Oxford.

Gallimard (n.d.): Catalogue Bande Dessinée. http://www.bd.gallimard.fr/ouvrage-A62803-aya_de_yopougon.html (accessed May 8, 2013).

KANNEH, K. (1995): Feminism and the Colonial Body. B. ASHCROFT, G. GRIFFITHS and H. TIFFIN (Eds.): *The Post-Colonial Studies Reader*. London, 346–348.

KELLY, K. (2000): *Performing Virginity and Testing Chastity in the Middle Ages*. London.

LAPIERE, D. and J.-P. STASSEN. (1999): *Le Bar du vieux Français* (Edition intégrale). Paris.

LEMAIRE, T. (2009): Thierry Groensteen: Les auteurs femmes sont très fréquemment réorientées vers l'illustration jeunesse. *Actua BD*. http://www.actuabd.com/Thierry-Groensteen-Les-auteurs (accessed September 3, 2012).

24 "travel tale"

25 "crime novel"

LOINGSIGH, A. (2003): Immigration, tourism and postcolonial reinventions of travel. C. FORSDICK and D. MURPHY (Eds.): *Francophone Postcolonial Studies.* London, 155–163.

MACLEOD, C. (2011): (Post)-colonial persistence: the presence of the "Black Venus" in the work of Warnauts and Raives. *Contemporary French Civilisation* 36 (3), 287–302.

MCKINNEY, M. (2000): The Representation of Ethnic Minority Women in Comic Books. J. FREEDMAN and C. TARR (Eds.): *Women, Immigration and Identities in France. Oxford:* Berg, 85–102.

WARNAUTS, E. and G. RAIVES (1994): *Equatoriales.* Paris.

WARNAUTS, E. and G. RAIVES (1996): *Lettres d'Outremer.* Tournai.

WEST-SOOBY, J. (2008): Nowhere is perfect. J. WEST-SOOBY (Ed.): *Nowhere is perfect: French and Francophone utopias/dystopias*. Newark, 1–10.

Chapter 5 | *Juliet J. Fall*

PUT YOUR BODY ON THE LINE: AUTOBIOGRAPHICAL COMICS, EMPATHY AND PLURIVOCALITY

I don't remember when exactly I read my first comic book, but I do remember exactly how liberated and subversive I felt as a result.
(SAID 2001, i)

We may think that circulating alternative images will rally resistance, but we have to remember that graphic depictions can sometimes do no more than sensationalize events. When that happens, we respond with outrage periodically, but that outrage is not transformed into a sustained political resistance. Is there another way to act upon the senses, or to act from them, that resists both sensationalism and episodic outrage at the limits on the visual imposed by techniques of war waging?
(BUTLER 2010, xiv)

POPULAR GEOPOLITICS AND THE POLITICS OF REPRESENTATION

It is a truism to say that we live in a specific political and historical context in which images, and their specific expressivity, are called into question in a number of ways (CAMPBELL 2003). In an introduction to Joe SACCO's autobiographical comic book *Palestine*, Edward SAID wrote that "as we also live in a media-saturated world in which a huge preponderance of the world's news images are controlled and diffused by a handful of men sitting in places like London and New York, a stream of comic book images and words, assertively etched, at times grotesquely emphatic and distended to match the extreme situations they depict, provide a remarkable antidote" (SAID 2001, iii). In this paper, I want to take this idea of comics as antidotes seriously, while remaining aware of the personal dimension of what SAID is talking about in the opening quote: that comics are read in particular places, and speak to us in very personal ways, feeding into and reworking our own personal histories. This could be connected to the literature on reading and the circulations of texts that has recently become of interest to geographers (KEIGHREN 2006; DITTMER 2010), as they have engaged not only with the strategic and situated production of knowledge, but also on its situated reception in multiple and uneven places.

That texts and images circulate is a given. Attempts to govern and control these circulations suffuse human history. Yet these are times when the specific control and dissemination of images of conflict – dead bodies, sites of bombing, atrocities – have taken on a new urgency. Marianne Hirsch, amongst many others, has written that countless new restrictions on images of war shape a new and urgent context for a sustained discussion of words and images, of reading and looking (WHITLOCK 2006, 965). As Judith BUTLER has noted, forms of dissent are the object of intense scrutiny in such contexts, as these are systematically excluded or controlled in order to maintain a climate of fear that serves the interests of the powerful. Likewise, in a discussion of Susan Sontag's work on the power of photographs to communicate the suffering of others, WHITLOCK argues that what is at stake are "fundamental questions about the interpretation of visual images and about their power to relay affect and invoke a moral and ethical responsiveness in the viewer regarding the suffering of others" (WHITLOCK 2006, 965). This is not just a theoretical debate: images can be deadly and dangerous, as during episodes of the so-called 'comic wars' when cartoons of Muhammad were instrumentalised in calling for violence against individuals in contexts where such images were considered sacrilegious (NORTON 2011). This led WHITLOCK and POLETTI to argue that attention to the specific productions, locations and relocations of the reproduction and consumption of comics "is a lesson of the controversies of 2005–2006, the so-called "cartoon wars": images travel far from their origins into very different communities of interpretation; their meanings are always contextual, social, cultural, and political as well as aesthetic" (WHITLOCK and POLETTI 2008, ix).

In this chapter, I aim to play with Judith BUTLER's proposal to put 'your body on the line', by considering three recent autobiographical comics that focus on political border issues and divisions in the Middle East. That comics deal quite literally and theoretically with drawn lines – *la ligne clair*, most notably – seems to make this image particularly apt.[1] Guy DELISLE's *Jerusalem*, Maximilien LE ROY's *Faire le Mur*, and Joe SACCO's *Palestine* all provide different alternatives to dominant geopolitical narratives. I have chosen these as three very different autobiographical books are all located in Israel / Palestine. They each provide on-the-ground accounts of daily life in situations of conflict, dwelling on the experiences of division and violence. All three are narrated by outsiders, yet make different use of local voices. It is this diverse experience of plurivocality within autobiographical comics that forms the main focus of this chapter. These three authors address explicitly the question of representation and image production and circulation, as they depict themselves and others sketching, taking photographs, reproducing and redrawing iconic images of conflict: literally framing their daily experiences visually.

My approach is sympathetic to PRATT's suggestion (2005) to use the materiality of the body as an opening for a politics of representation, making space for a non-essentialist 'politics on the ground', constituting new subjects. The daily geopolitics of divided spaces is explored through these tales of personal experience as narrated and encountered in autobiographical comics, drawing attention to the

---

1 I thank Jason DITTMER for pointing this out – as my original title was a completely unnoticed allusion to this!

shared yet diversely experienced violence of boundaries as multiple material and political objects. Through this appeal to intrinsically visual texts that play on past memories of particular places yet assign universal meanings to the absurdity of division, I discuss the possibilities for developing graphical alternatives to the geopolitical 'view from nowhere'. I aim to explore one visual medium for developing what BUTLER calls an "anti-war politics that focuses on the dispossessed and those rendered precarious in ways that require new vocabularies and new practices" (BUTLER 2010, x). Her work suggests questioning the framing of war directly, since "the frame does not simply exhibit reality, but actively participates in a strategy of containment, selectively producing and enforcing what will count as reality" (BUTLER 2010, xiii), instrumentalizing certain versions of reality. I want to suggest that comics can frame – quite literally, within boxes – alternative and polyvocal geopolitical visions.

QUESTIONING THE GEOPOLITICAL GAZE: RETHINKING REPRESENTATION

Within geography, there have been sustained feminist critiques of the specific viewpoint adopted by critical geopolitics (DOWLER and SHARP 2001; HYNDMAN 2004; SHARP 2007). These proposals resonated within the discipline, calling for ground-level, local and embodied sites of moral propinquity or nearness ("direct, personal, honest and angry" – TOAL 1996, 173). HYNDMAN's invitation to develop a strong feminist geopolitics "to extend the work of arguably disembodied critical geopolitical analysis by (re)situating knowledge production as a partial view from somewhere" (HYNDMAN 2004, 309) made a lasting impression on many. In a similar creative and critical vein, an increased focus on popular geopolitics – i.e. on non-traditional, everyday discourses such as those of popular culture including film, fiction and (to a lesser extent) comics – has become apparent within geography, perhaps finding new legitimacy in this new appellation, distancing itself from Culture or Media Studies while drawing heavily from such approaches. But if we follow HOLLAND this "has not adequately considered how the formats it interrogates are simultaneously appropriated as formats of resistance" (HOLLAND 2012, 107). The role played by visuality is centrally identified as problematic in so-called traditional or realist geopolitics yet this is not interrogated adequately as problematic in the emerging counter-geopolitical work. That such productions of unofficial (i.e. non-state) art are necessarily and intrinsically oppositional and/or resistant should not however be taken for granted, as CAMPBELL (2003) has pointed out in his discussion of cultural governance, drawing on Michael Shapiro's discussion of the term. KUUS notes that a key challenge is to avoid glamorizing resistance and civil society in general: to show the diversity of resistance, the entanglements of domination and resistance, and the futility of looking for the 'self-evidently good' (KUUS 2010). She suggests instead that attention to passiveness, irony and anonymity is more fruitful. Taking the central feminist tenet that 'the personal is political' seriously, such work has argued that the personal is also geopolitical, leading to a scholarly focus on everyday political practice and more embodied accounts of power.

The following page taken from Joe SACCO's *Palestine* is a good entry point: a short episode, almost a parenthesis, that illustrates how daily lives are transformed by division (Figure 5.1). As the narrator and his friend walk through a fenced, violent landscape, the daily strategies of survival are suggested in all their pathetic ordinariness. A young girl, on her way home from school, takes a shortcut through a fence: a small act of resistance, no doubt taking absurd risks to shorten her walk, or else simply to make it possible. That this is barely commented upon by the two men does much to show how such practices have become common place, taken for granted, banal. Yet this, of course, is staged indifference: the narrator has very carefully observed the scene, or at least enough to draw it in dramatic detail. The apparent nonchalance is a clever trick. Death-defying actions have become commonplace, vulnerability absolute.

So how can we make sense of these depictions of embodied, quotidian experiences of military power that although narrated by Western observers appear to capture and transmit something of the personal experiences of violence? Is this just a clever narrative trick, or could it be politically subversive? And what must we make of the positionality of the authors of many such comics – Western quasi-elites (although, by Western standards, I suspect you have to be pretty famous as a comic artist to consider yourself realistically as part of any social elite – able to extract themselves from such situations of violence when they choose?

Although geographers have repeatedly called for such decentred critical geopolitical accounts, what this translates into in practice is not always clear in many such scholarly contributions, and the field is undoubtedly still seeking coherence. NICLEY goes some way in suggesting what this might look like, arguing that we need a focus on "situated, meaning-laden practices of place-bound individuals heterogeneously constituted through racial, ethnic, class, gender and other identities (…) without lapsing into an uncritical reification of the place concept" (NICLEY 2009, 20). He calls for "a commitment to thick description of the complex, hybrid geographies through which individuals construct their everyday lives (…) to unearth the causal processes that bind elite geopolitical language to the ways that people adopt, internalize and challenge such explanations" (NICLEY 2009, 21). NICLEY's, like HYNDMAN's (2004), is very much an agenda-setting position paper, with no clear method other than an indication that ethnography might provide useful material. Yet his central call to listen "to the voices of the marginalized and to achieve a greater sense of shared understanding of both our commonalities and differences" (NICLEY 2009, 22) is poignant, and speaks to what many comic book authors are apparently trying to do when providing alternative political narratives that focus on individuals, both Western travellers and local people, in conflict situations. But how can we make sense of these polyvocal graphical geopolitical narratives?

DELISLE frequently uses this explicit staging of his own observation of daily situations as a *mise-en-abîme* of his own subsequent image production, simultaneously drawing attention in this vignette to the plurivocality – and ubiquity – of this process of visualizing and recording daily experiences. Politics, here, is also a performance, and recognized as such.

AUTOBIOGRAPHICAL COMICS

WHITLOCK suggested coining the term "autographics" (WHITLOCK 2006; WHITLOCK and POLETTI 2008) for the specific comic genre of autobiographies presented as graphic memoirs in which the subject position navigates through different representations of selfhood, addressing truth and identity in a variety of changing discourses. Here, I am particularly interested in how autobiographical avatars engage actively with the conventions of comics, making us question the genre as self-reflective authors play with form, in particular by stepping outside of the gutters and frames of comics. Sorting selfhood and grief quite literally 'into boxes', as SPIEGELMAN has suggested (quoted in WHITLOCK 2006, 968), produces a unique aesthetics that combines drawing, design and writing, in particular within the self-regarding art of graphic autobiography. This creativity is particularly useful in presenting a traumatic, situated side of history, while refusing to "show it through the lens of unspeakability or invisibility, instead registering its difficulty through inventive (and various) textual practice" (CHUTE 2008 in HOLLAND 2012, 107).

As I have written elsewhere, bodies themselves are objects of resistance – and bodies represented in comics at times specifically so (FALL 2006). Bodies need not be subjugated to coercive power but can in themselves be objects of resistance – and particularly when they are very directly representing themselves in these autobiographical narratives. Following BUTLER, it is useful to remember that the body is fluid, not fixed, allowing for "a notion of matter, not as a site or surface, but as a process of materialization that stabilizes over time to produce the effect of boundary, fixity, and surface we call matter" (BUTLER 1993, 9). Sites of performance in their own right, rather than surfaces for discursive inscription, bodies participate actively in their own making. Adding this layer of agency and potentials for resistance is particularly important in the case of autographics: the author, through their autobiographical avatar, is a particularly active, visible participant in crafting its own selfhood. Before being authors, these artists were witnesses, travellers, recorders and – in some cases – professional journalists, occupying what FARISH has called an interstitial position in conflict situations, "working between embodied proximity and disembodied detachment" (FARISH 2001 in HOLLAND 2012, 119), emplacing moral responsibility at certain geographic locations through their rescripting of their lived or narrated tales. This agency in scripting their own stories is crucial in avoiding the pitfall of revictimization by deploying "victims as an antidote to heroic images" (CAMPBELL 2003, 67), which risks producing a generalized and standardized visual account that anonymises victims and depoliticizes conflict.

But authors in this genre are also plural, and autographics are multi-layered. MILLER raises this in arguing that "the reportage is made more effective by the capacity of *bande dessinée* for plurivocality as well as for the plurality of ways in which narratorial intervention can be made apparent. Ultimately, it is the essence of a medium that does not solely depend on mechanical reproduction but is mediated by the artists' hands and eyes that accounts for the impact of this highly nuanced and personal portrayal of the political process" (MILLER 2008, 115). I will return to this point in particular when discussing the specific plurivocality of LE ROY's *Faire le Mur*, in particular regarding the gradations in detachment and subjectivity that this allows.

PICTURING EMPATHY

WHITLOCK (2006) mentions the importance of thinking about the pleasures and pains of reading autographics, as evidenced in the opening quote of this paper by SAID (2001). She further notes that several authors have confirmed the potential in comics for distinctive mediations of trauma and cultural difference and for innovations that open up some new ways of thinking about the ethics of life narrative as they move across cultures. But what has been called the 'comics wars' – and the deadly impact of specific comics in contexts beyond their production – needs to make us wary of assigning them universal value. "The cartoon wars indicate that graphic art moves as a commodity in a global market across various econo-, ethno-, and ideoscapes, but at the same time they are a cautionary reminder that difference is not transcended or resolved in these transits, and visual images are processed within vastly different communities of interpretation, and easily coopted as propaganda" (WHITLOCK 2006, 970).

Many analyses are however subtler in considering cultural issues. Discussing Marjane SATRAPI's marvelous *Persepolis* (2000), WHITLOCK notes how an active mediation of cross-cultural relations occurs unambiguously in certain autographics, "drawing on the capacity of comics to free us to think and imagine differently in times of trauma and censorship" (WHITLOCK 2006, 973), opening up new and troubled spaces of representation. In cautious attention to detail, such as in SATRAPI's subtle discussion of the introduction of the veil in post-revolutionary Iran, "we can begin to recognize the norms that govern which lives will be regarded as human and the frames through which discourse and visual representation proceed" (WHITLOCK 2006, 976). Yet we must be cautious of suggesting – as MCCLOUD has done (2006) – that comics are universal[2], even when "they seem to acquire a life of their own, with recurring characters, plot situations, and phrases

2 WHITLOCK and POLETTI would however run to MCCLOUD's defence, arguing plausibly that his world-vision is much wider since "although his avatar appears to be comfortably established in middle America and introduces himself with a chummy "Hi. I'm Scott," his "nation" of comics is cosmopolitan and intertextual, a global industry shaped by Tezuka, Herge, Miyazaki, Tardi, Doucet – the list is extensive. Along with Will Eisner, Charles Hatfield, and R. C. Harvey, among others, McCloud insists on reading comics as a rich global cultural formation" (WHITLOCK and POLETTI 2008, xii).

that turn their readers, whether in Egypt, India or Canada, into a sort of club in which every member knows and can refer to a whole set of common assumptions and names" (SAID 2001, i). Instead, I would follow WHITLOCK's greater nuance in arguing that "there can be no simple universality in the associations produced by cartooning across very different relationships. (…) We find this elsewhere in representations of trauma in autographics, where troubled memories are held in the boxes of grief of sequential art" (WHITLOCK 2006, 977). She suggests further that comics free us to think, imagine and see differently, engaging with the pain and suffering of others, offering a clear link to what I mentioned earlier was a central ambition of feminist geopolitics.

The concept of empathy is crucial to understanding how such comics engage the reader. If empathy is grounded "not in affinity (*feeling for* another insofar as we can imagine *being* that other) but on a *feeling for* another that entails an encounter with something irreducible and different, often inaccessible" (BENNETT 2005, 10), then authors, as witnesses to particular situations, are indeed engaged in a form of "empathetic witnessing." In her book on empathic vision, trauma and art, BENNETT argues that by combining affective and intellectual operations, empathy can actually be considered a mode of seeing (BENNETT 2005), a particularly visual way of conceptualising it that is apt here. This requires recognizing an alterity, irreducible to our own experience[3]. However, while "empathetic witnessing" has been used to describe particular forms of photo reportage, using it to describe a drawn renditions of real situations in comic form could be more problematic, nothwithstanding the crafting and politics also involved in photography.

A play on the format of comics as predominantly associated with fiction can also be strategically played up, as SACCO does in his book *Palestine* in noting that "Make no mistake, everywhere you go, not just in Marvel Comics, there's parallel universes" (SACCO 2001, 102). As in his other numerous works (see HOLLAND, *this volume*), SACCO engages the reader by contrasting his often-unflattering naïveté and unease to the traumatic experiences recounted by those around him. Drawing upon an encounter with Jabril, a Palestinian living in Balata refugee camp, he recounts a very corporeal attempt to empathize with his interview partners. In detailing the experience of torture of another, recounted in a "middle class living room in East Jerusalem" (SACCO 2001, 102), a banal space of daily life and hospitality infused with the daily rituals of tea and snacks, he writes that: "I want to hear about his two months' grilling in Nablus prison, the time the Israelis accused him of training with the Popular Front" (SACCO 2001, 93). As they speak, they are joined by Abu Akram, another contact of SACCO's, who – when asked whether he has ever been interrogated – proceeds to engage in direct role play, placing SACCO in the position of detainee (Figure 5.3).

3 To put it in simpler terms, this means precisely not assuming that we can comprehend the suffering of others, or ever liken it to our own (i.e. never saying, to someone in pain, that "I know just how you feel").

Despite his unease with this *mise-en-scène* of torture, SACCO cannot help but continue to pursue his attempt to experience empathy physically by requesting more 'itsy-bitsy details', more concrete material and narrative elements to make bodily sense of the unreality of another's experience of violence, in a setting returned to the banalities of ordinary life: a family sitting room, where tea is served. This clear staging of the experience of empathy resonates with the reader's own attempt to make sense of what is going on, as we follow the narrator through many similar interviews. That SACCO positions himself here, as elsewhere, as having been "raised a suburban schoolboy..." (SACCO 2001, 94), a clear outsider stripped of points of reference, and thereby making explicit how alien and how disembodied this experience of extreme violence is to him, helps the reader to align with his position. "Me?", he writes, "I wonder how long I'd last getting the business behind a closed door... Not long, I bet, but I'm a Pussy First Class... a harsh word and a dirty look and I'd be screaming for Amnesty Int'l" (SACCO 2001, 97). Yet despite this constant distancing, the connections with people he weaves are effective in showing just how suffused with violence and absurd rules daily lives have become, and how mental toughness, ability to withstand humiliation, and cope through cynical and absurd humour become the only true measures of valor.

Because violence for SACCO is always experienced vicariously, this appears to reinforce his position as an outsider, and yet at the same time this also grounds his legitimacy as an outside observer. The reader is therefore empathizing both with SACCO's 'once removed' experience of violence, and with the more immediate experiences of his characters, all the while remaining more or less aware that these are mediated and narrated by him. The seamlessness with which these points of view – this plurivocality – is staged is remarkable, made all the more concrete by SACCO appearing as a character in the images, one among many, narrator, actor, producer. That this constant de/recentering is staged visually removes all feeling that this is simply a God-trick, a point of view from nowhere.

Drawing on personal, mundane details to establish moral proximity and embed places and people with our own everyday experience is highly effective, as TOAL has noted in another context (1996, 176). Indeed the focus in many autographics is on the everyday, on "history's losers, banished to the fringes where they seem so despondently to loiter, without much hope or organization, except their sheer indomitability, their mostly unspoken will to go on, and their willingness to cling to their story, to retell it, and to resist designs to sweep them away altogether" (SAID 2001, v). There is however a paradox here: in the comic SAID is referring to, as in the others discussed here, it is not the people themselves writing the book, but particular authors performing a form of reported speech, and a cleverly-crafted one at that. That many of these authors are part of a global elite of Western travellers or reporters must lead us to exercise caution in assuming these articulate or reproduce the voices of the voiceless. This is not, at least in the cases considered here, a 'subaltern' form of politics (SHARP 2011), despite being potentially emancipatory. Instead, it seems to me, their tremendous strength lies in their ability to give voice to what at least appears to be a plurality of positions, a plurivocality of daily experiences of violence – including that of the author as both observer and narrator.

In many ways, it is precisely because their experiences as travellers mirror our own experiences of bafflement and unease when travelling to or living in new locations that their own empathy touches us so deeply (and by this I mean the 'Western readers' at whom these books are largely aimed). If it is of course impossible to anticipate how such comics will be received, as this is never straightforward, this does not mean that the authors do not have clarity of purpose when crafting them. These forms of post-reportage, speaking in "considered retrospect" (WALKER 1995 in CAMPBELL 2003, 73), can specifically highlight the often contradictory narratives of particular situations, serving to destabilize hegemonic or simplistic narratives by focusing on the vulnerability and individual suffering of those involved, specifically through the use of multiple voices and points of view. The gutters of comics, in particular, have been likened to aporia, or "blank spaces where new meanings can be generated and a distinctive cross-cultural translation can occur" (WHITLOCK 2006, 978).

SAMSON goes further in analyzing this ability of comics to move readers, suggesting that paradoxically by their very nature war and conflicts crush the individual, potentially providing scant detailed material for telling stories, other than grandiose – and depersonalized – tales of heroism. Yet the universality of the human experience of fragility means that, if well rendered, we cannot be indifferent to the complex material emerging from situations of conflict:

> By crushing the imagination and conscience of those who pay tribute to it, war works to destroy individuality. That is if the singularity of individual experience hasn't already been crushed by grandiose myths and idealized narratives. That would be after all just another way of conflating the lives of ordinary civilians, making them nameless and faceless. […] If we consider war in its most singular or in its collective dimensions, then war can only show us how fragile human lives are, how close to meltdown humanity can get. We are by its very nature all touched by it, relating as it does to our own human experience. It is our concern, even in peacetime and even when we haven't experienced it closely[4] (SAMSON 2011, 255).

This reflexive experience of fragility and humanity can be made into a form that bridges distance, and engages readers. SAMSON argues that authors in this genre are able to build on deeply-held desires to know more about the people they meet, the situations and events they find themselves in. He suggests that this is not a search for forms of institutionalized knowledge that can be more or less removed from daily experiences on the ground, but rather a quest for forms of knowledge

4 Own translation from: "En broyant l'imaginaire et la conscience de ceux qui lui paient un tribut, la guerre travaille à l'anéantissement de l'individualité. A moins qu'elle ne dissolve d'emblée cette singularité sur l'autel de mythes et idéaux intégrateurs. Ce qui est une autre façon de niveler la mêlée des sans-grades, sans-nom et sans-visage. (…) Qu'il s'agisse de l'envisager dans ses aspects les plus individuels ou les plus collectifs, la guerre ne montre jamais qu'une humanité fragilisée, en pleine débâcle. En cela, elle nous concerne tous, autant que nous sommes. Elle est notre affaire, même en temps de paix et même quand on ne l'a pas éprouvée de près."

that bring him closer to grasping the complexities of the world in so far as these are experienced and felt by those he meets. "We are dealing with knowledge linked to life experience and that implies, because of that, a transitive dimension"[5] (SAMSON 2011, 260). SAMSON argues that this is made possible by the depth of narrative that emerges from the carefully-built relationships of friendship and mutual respect between an author and his/her subjects. This ability to provoke real empathy is then passed on to the reader who can identify intimately with what it is to be human – rather than with grand historical narratives. The challenge, then, is to find an appropriate format to render this introspection and complexity graphically. It seems to me that this is very close to what SACCO explains as his project:

> If you have to say what my motives for doing it are, what I'm trying to get people to get out of it, I'm trying to show a story of a town in a war in an artistic way. Because I think a lot of people are just totally turned off by these sorts of things. They feel that this sort of topic is beyond their comprehension or that it's going to be too much to take. So the idea was to convey something that, at its heart, was journalistic. But I wanted to convey it in a way that would bring out the humanity of the people there. [...] And, yeah, you want people to be moved by it [...] I'm interested in informing, but that's nothing compared to the impact I want to have [...]. You should read about it because you care about it (SACCO 2011, 254).

EMPATHY AND ALIENATION: MAKING SENSE OF THE MUNDANE

Guy DELISLE's tale of expat life in Jerusalem develops a sense of personal alienation associated with the need to combine a quasi-daily experience of violence with the rhythms of daily life. This is made even stronger through his apparent solitude in many of the images: appearing alone, or with a few chosen recurrent characters he is shown progressively making friends with. DELISLE's position as a detached observer plays explicitly on his role as mediator, exploring and recounting his travels: he is the trailing spouse of a logistics specialist with *Médecins Sans Frontières*, an international medical charity, spending a year in Jerusalem running the household during what turns into a sort of sabbatical (DELISLE 2011, 226). In narrating and illustrating his own explorations of the city and the wider region as a detached and explicitly marginal participant, he draws the reader into empathizing with him simultaneously as a parent of two young children, a spouse, and a traveller. It is only towards the end of his tale that he reasserts his own professional identity, literally finding his own space within an improbably charming work studio attached to a church. DELISLE illustrates the daily rhythms of his new life (Figure 5.4), since in addition to the many narrated perambulations and explorations of the city, most of his time is spent as "the housewife" (*la femme au foyer* – his term) taking care of family logistics. These are given particular poignancy by the local geopolitical context that has a specific impact on the family's expatriate daily life: his two young children attend daycare centres run respectively along

---

5 Own translation from: "On parle d'un savoir lié à l'expérience de la vie et comportant, de ce fait, une dimension de transitivité."

Christian / Jewish and Christian / Jewish / Moslem weekly calendars, implying different 'weekend' breaks, while his wife has Fridays and Saturdays off. Particular events, in this extract Ramadan, further throw off the carefully-honed routines. The personal is inescapably (geo)political. His simple tabular illustration of this, with his own weekly routine having no particular pattern at all – always on, yet coming last, as many full-time parents could recognize – is a clever way of connecting the personal with the wider political context. His own exhaustion at the end of the day, with no energy to play, does much to show that this is not a life of leisure, or the carefree life of an artist on sabbatical.

These careful vignettes of daily life are clever strategies for giving depth to the characters, and to DELISLE in particular. We can empathize with his daily struggles, and what can be the grinding boredom of daily family routines. This works as a strong counterpoint to life outside the home, serving to offset the stark violence of division on the ground. It is here that his pure and almost naïve style of drawing serves him particularly well: the sheer size and volume of the Wall being built by the state of Israel fills frame after frame, with the narrator dwarfed by its sheer scale. Referring again to the weekly rhythms of violence – many Muslims wish to cross the checkpoints to go and pray in the Great Mosque on Fridays – he recounts a trip to Qalandiya with an international observer, and his first view of the Wall (Figure 5.5). "*I had never seen the separation wall. I didn't think it was that high*", he says, in his narrator's voice. Yet, in the frame, all he finds to say is "*Eh ben!*", a trivial exclamation similar to a matter-of-fact, understated "*Wow!*" For how can any outsider make sense of this spatial expression of territorial violence, conjuring up – for a Westerner – other half-forgotten images of division? This clever use of his own decentered plurivocality and his own slow attempts to make sense of the violence of such impositions – between the moment of encounter and that of drawing – are cleverly and carefully rendered in words and images. The immediacy of the encounter, the subsequent narration of his reactions as a memory, together with the carefully-chosen framing of the moment that plays on the crushing perspective viewed from the ground, builds up a geopolitical landscape built on daily encounters, lived experience and no longer solely on geopolitical maps.

In this alien landscape, the threat of violence is never far ("*On va laisser la voiture ici. Je préfère qu'elle soit loin. On ne sait jamais*" – "We'll leave the car here. I prefer it to be far away. You never know"). Yet DELISLE makes clear that he stands outside this violence, he is just an observer, and an awkward and almost illegitimate one at that, as he wears the supposedly protective jacket of the observer-organization his guide belongs to.

This simultaneity of the production of images was illustrated earlier in Figure 5.2 where the taking and recording of images seemed a bizarre and integral part of the whole performance of violence itself. For, unlike in SACCO's world, where violence is narrated and depicted repeatedly, and restaged for effect as in Figure 5.3, DELISLE's experience of the violence of the geopolitical situation is more muted, but no less powerful in provoking intense empathy. Instead, his understated daily experiences are grounded in almost-permanent spatial confusion. He spends his time mostly lost, disoriented, building up his own mental-map of the city as a divided space of multi-layered alienation and separated lives. Jerusalem only begins to make sense to him as he slowly connects its unmapped spaces to his own routines, new friends and simple daily habits: playgrounds for his children to play in, cafés to sit in peacefully, and finally his own workspace to think, write and draw in. If at the time, as he makes clear, he had expected to be finishing a book on Burma, not the Middle East, he ends up only sketching his daily experiences. The book itself was formalised subsequently, and he only contributed a few sketches to a blog during his time in Jerusalem, too involved in making sense of daily life to think of elsewhere.

Because his is a widely-shared experience of making sense of a new place, any new place, it leads to an easy empathy with the narrator. In a sense, this makes it less easy to connect with the other characters that appear in the book, appearing on the periphery yet giving body to the narrative. This apparent marginalization of others does not however make it any less powerful politically: DELISLE makes it clear that these are real people within his own biographical tale, individually occupied by making sense of the violent background in their own daily lives and routines, each finding space to exist in the absurdity. He makes no claims to speak on their behalf, and provides only scant background to each. This is understated empathetic witnessing, focusing on daily struggles, creating alternative narratives that directly connect with readers. It is, however, the tale of a baffled outsider, and unashamedly presented as such. The boundaries, walls and checkpoints he encounters are inconvenient, surprising, and absurd, but he is not threatened or made especially vulnerable by them – at most, these make him lose time. Instead, they are almost like a hidden code, a spatial language of territorial violence that has to learn to be able to maintain his mobility. It is clear that, at the end of the year, he can go home. This, however, does not make his portrayal of the daily strategies of others any less potent. On the contrary, these vignettes of human ingenuity, that depict others more adept at functioning within these bounded spaces than him, foil what could have been construed as an attempt to appropriate or assimilate their experiences to his own, avoiding what BENNETT (2005) calls 'crude empathy'.

REPRESENTING PLURIVOCALITY GRAPHICALLY

In contrast to this detached yet also touching viewpoint, I want to spend the rest of this chapter discussing another more complex standpoint: LE ROY's *Faire le Mur* that plays explicitly with the plurivocality that comics can develop graphically, through the use of multiple authors. Here again, I focus on the effects and impacts of division that are experienced differently, and how these are framed and represented. The title of the book is worth unpacking for an Anglophone audience: while *faire le mur* can literally mean 'making the wall' it also, figuratively, means escaping from somewhere undetected and unauthorised, an apt image in this case.

The book is the story of several encounters. It emerged from the friendship that binds Maximilien LE ROY, the author, with Mahmoud Abu Srour, a Palestinian artist of the same age living in the Aïda refugee camp in July 2008. The narrative is woven around the unrequited love of Mahmoud for Audrey, one of the many Western women to pass through the camp, and who inevitably leaves. Lastly, it reflects the encounter between LE ROY and the Palestinian experience of confinement and bounding. The latter is viewed empathetically through another's eyes: Mahmoud as his peer, his soulmate. What is unique however is how far LE ROY goes in playing explicitly on the plurivocality of the narrative: it is not only him narrating the tale – the biography – of his friend Mahmoud in his main character's voice, since both provide starkly-different images throughout the book. Mahmoud, perhaps rather unsettlingly, does not however appear as co-author. Yet this is believably a story in two voices, made all the more striking by their different graphical styles and use of colour.

This book combines multiple media: the narrative itself, and three illustrative appendices that give body to the political context, providing background 'facts' to a comic format usually associated with fiction. The first annex is made to look like a family photo album, showing the real-life family of Mahmoud, his former childhood home and his now-occupied farmland, moments of past innocence and happiness in stark contrast to his present life in the refugee camp that is not photographed, only drawn. The second is a photo-reportage of iconic Israeli landscapes, including well-known graffiti on the Wall and photos of Palestinian children both in daily routines (on the school bus, in the street) and situations of violence (throwing stones at soldiers), and photos of checkpoints, destroyed buildings and gleamingly-new Israeli settlements. The third annexe is an interview with Alain Gresh, an internationalist thinker, apparently carried out by the author, and rather inexplicably but plausibly offered to provide a political alternative to narratives of inevitable ethnic divisions and identities.

The main body of the book is the graphic biography of Mahmoud[6], flitting between three different times: the present and the soulless boredom of his daily life selling wares in his parents' corner shop; the distant past and his return to Palestine after years working in Israel to build a house, with its subsequent destruction and replacement with an Israeli settlement; and the near-past with the non-event of his failed seduction of Audrey (Figures 5.6, 5.7 and 5.8). Unlike in the two books discussed earlier, LE ROY himself never features in the comic. The voiceover is meant to be Mahmoud himself narrating his own tale: this is not an autographic in the sense discussed earlier, but rather a polyvocal biography.

As the narrative flits between three epochs, drawn alternately by both authors, it provides a very direct visual experience of plurivocality. Mahmoud's tortured, and colourful, retelling of the moment of exile, when his house was smashed up by Israeli forces to make way for a settlement, and the trickery involved in this as they politely detained his father and him for an apparent 'simple formality' while

6 I use his first name here in the discussion, echoing LE ROY's use.

in fact crushing their home, contrasts starkly with the dour interiors of his subsequent life in the refugee camp as drawn by LE ROY. In Mahmoud's images, LE ROY still provides the narrative, composing and captioning the images himself as main author.

The differently experienced effects of boundaries features highly in the story: when Audrey requests to go and see some 'real Palestinians', he decides to take her to his family land, on the other side of the Wall. He hopes this will lead to a romantic encounter. While she can cross without trouble, holding an international (French?) passport, he has to risk detainment by crossing undetected, running across a wasteland and climbing through a fence. Their encounter is presented as impossible, and doomed: she simply has no idea of his experiences, or the risks he has taken to get there, and flippantly decides to go back on the first day. She is apparently oblivious to Mahmoud's attraction and the sacrifices made to be there: two lives in different worlds. When they briefly meet up again, several weeks later, she briefly asks him how his return was. He got caught by soldiers, and was detained. She lives in a political geometry where boundaries are mere inconveniences (Figure 5.9).

The image of the bird, a clear signalling of Mahmoud's own caging, contrasts with the apparent triviality of the dialogue. Mahmoud realises it is pointless to try to explain what happened to him to Audrey, and in any case she is leaving the country.

These alienated souls, caged in different worlds, cannot have any meaningful connection: Audrey through her apparent inability to connect to Mahmoud or to his experience; and he through his position as an exiled Palestinian seeking escape through his art while physically bounded by political divisions. She, like DELISLE and SACCO discussed earlier, is part of the global elite, endlessly mobile, just passing through these landscapes of violence. Unlike them, however, she appears unaffected by it, or simply ignorant. Yet this is only Mahmoud's/LE ROY's viewpoint: we could suppose that she might be understandably cautious of a romantic entanglement in such a highly-charged setting. In any case, the author is not trying to foster empathy with her: her other-worldly, haunting appearance within Mahmoud's drawings reduce her to an icon, a dream of escape, while simultaneously cast as a body desired, lovingly and repeatedly drawn. An unattainable muse, on the other side of a fence.

I argued earlier that comics can highlight the often contradictory narratives of particular situations, destabilizing hegemonic and simplistic narratives by focusing on the vulnerability of individuals suffering, specifically through the use of multiple voices and points of view. It seems to me that this particular book does this admirably, while maintaining unity of narrative: Mahmoud, although 'written about', is given his own voice through his drawings, as LE ROY connects empathically with him as a young man and artist of his own age, but one stuck in an impossible political situation. LE ROY is thus a witness and invisible narrator, but narrowly avoids the trap of 'speaking for' his main protagonist. Grief, as Spiegelman was quoted as saying earlier, is indeed sorted into boxes in this book – on the page, of course, but also graphically in drawn bird cages, walled interiors and scarred landscapes – allowing the reader to encounter the mad, bad, sad geopolitics of the Middle East through diverse but specific viewpoints, connecting to the daily experiences of particular people.

CONCLUSIONS

The genre of autobiographical reportage comics has become a phenomenon (MILLER 2008). Comics are a fantastically creative format in their use of drawings, photos, drawings and animated or semi-animated images, reflecting constant experimental innovations (see for example CHAPPATTE's recent online use of semi-animation; the combination of photos and drawings in GUIBERT, LEFÈVRE and LEMERCIER's in *Le Photographe* (2003–2006); the use of photographs and drawings in SACCO's essay published in the special edition of *Safe Area Gorazde* (2011) that give fascinating insights into his reportage and writing processes; or KUBERT's *Fax from Sarajevo* (1996) that mixes photos and real-life documents and correspondence). The use of new media, in particular within websites and author blogs, has also provided new sites for additional graphic innovation, further changing the landscape of creation and diffusion of comics. In many such narratives, moments of historical interlude, flashbacks, reproductions of iconic images or photographs or simply banal everyday situations, are rendered subtly absurd through their retelling, and play on deeper and multiple meanings to create productive counter-hegemonic readings of geopolitical events. In this paper, I have considered three (auto)biographical comics set in Israel / Palestine, written by Western authors, arguing that the potential for plurivocal narrative offers a useful format for counter-geopolitical narratives that make use of empathic witnessing.

I have suggested that the three authors' aims of creating multiple, yet legible, stories that focus on encounters with marginalized, individualized characters, and the authors' slow integration and absorption into the local political contexts, are effective strategies for making the experience of marginalization and violence legible across contexts. In these oppositional works, as HOLLAND has written, "there is no predictable storyline, relying on pre-established histories, backgrounds or character tropes. More important, however, is the blending of complex subject and diverse narrative practice" (HOLLAND 2012, 114), unsettling conventional and popular portrayals of the experiences of conflict. BUTLER (2010) was quoted at the outset of this chapter, writing about the powers of images to act upon the senses. She is concerned by how the framing of war by the media, through and within key images and events, enshrines certain populations as more or less grievable. In battling the key role of images in such framing, she wonders whether the visual could also become the field in which we are solicited to assume responsibility to resist unjust war and the affirm convergent precarious conditions. In this chapter, I hope to have shown how comics can provide counter-frames of geopolitical conflicts that bring distant others into view. In a format that avoids conflict fatigue, comics can depict and make visible the vulnerability of individuals, focussing on their agency and autonomy beyond their status of victims, and can move beyond simply taking note of their precariousness to provide instead a political basis for critical outrage grounded in empathy. That they are also beautiful, aesthetically pleasing objects makes them all the more potent.

ACKNOWLEDGEMENTS

Earlier versions of this chapter were presented at the Borderscape III Conference in Trieste in June 2012, and the Geography Lectures at the University of Wageningen in February 2013. Special thanks to Jason Dittmer for his ever-careful reading, encouragement and excellent suggestions on this manuscript, and to Jean-François Staszak, Claudio Minca and Michael Shapiro for earlier very useful suggestions.

REFERENCES

BENNETT, J. (2005) *Empathic Vision: Affect, Trauma and Contemporary Art.* Stanford.

BUTLER, J. (1993) *Bodies that Matter: On the Discursive Limits of Sex.* New York.

BUTLER, J. (2010) *Frames of War: When is Life Grievable?* New York.

CAMPBELL, D. (2003): Cultural governance and pictorial resistance: reflections on the imaging of war. *Review of International Studies* 29, 57–73.

CAMPBELL, D. (2007) Geopolitics and visuality: sighting the Darfur conflict. *Political Geography* 26 (4), 357–382.

CHUTE, H. (2008) Comics as literature? Reading graphic narrative. *Publications of the Modern Language Association of America* 123 (2), 452–465.

DELISLE, G. (2011): *Chroniques de Jérusalem.* Paris.

DITTMER, J. (2010): Comic book visualities: a methodological manifesto on geography, montage and narration. *Transactions of the Institute of British Geographers* 35(2), 222–236.

DOWLER, L. and J. SHARP (2001): A feminist geopolitics? *Space and Polity.* 5(3), 165–176.

FALL, J. J. (2006): Embodied geographies: naturalised boundaries and *un*critical geopolitics in Schuiten and Peeters' 'La Frontière Invisible'. *Environment and Planning D: Society and Space* 24 (5), 652–669.

FARISH, M. (2001): Modern witnesses: foreign correspondents, geopolitical vision, and the First World War. *Transactions of the Institute of British Geographers* 26 (3), 273–287.

FOLMAN, A. and D. POLONSKY (2009): *Waltz with Bashir: A Lebanon War Story.* Atlantic Books.

GUIBERT, E., D. LEFEVRE and D. LEFEVRE (2003, 2004, 2006): *Le Photographe.* Paris.

HOLLAND, E. C. (2012): To think and imagine and see differently: popular geopolitics, graphic narrative, and Joe Sacco's "Chechen War, Chechen Women." *Geopolitics* 17(1), 105–129.

HUGHES, R. (2007): Through the looking blast: geopolitics and visual culture. *Geography Compass* 1(5), 976–994.

HYNDMAN, J. (2004): Mind the gap: bridging feminist and political geography through geopolitics. *Political Geography* 23, 307–322.

KEIGHREN I. M. (2006): Bringing geography to the book: charting the reception of "Influences of Geographic Environment." *Transactions of the Institute of British Geographers* 31, 525–540.

KUBERT, J. (1996): *Fax from Sarajevo: a Story of Survival.* Milwaukee.

KUUS, M. (2010): Critical Geopolitics. *The International Studies Encyclopedia* Volume II. R. Denemark. Oxford, 683–701.
LE ROY, M. (2010): *Faire le Mur.* Paris.
MARION, P. (2011): La guerre prise de vue: de Shooting War au Photographe. V. ALARY and B. MITAINE (Eds.): *Lignes de front: bande dessinée et totalitarisme.* Chêne-Bourg, 287–312.
MCCLOUD, S. (2006 (2007 for French edition)): *Faire de la Bande Dessinée.* Paris.
MCKINNEY, M. (2008): *History and Politics in French-language Comics and Graphic Novels.* Jackson.
MILLER, A. (2008): Citizenship and city spaces: bande dessinée as reportage. M. MCKINNEY (Ed.): *History and Politics in French-language Comics and Graphic Novels.* Jackson, 97–116.
NICLEY, E. P. (2009): Placing blame or blaming place? Embodiment, place and materiality in critical geopolitics. *Political Geography* 28, 19–22.
NORTON, A. (2011): The Danish Cartoon Controversy and the Challenges of Multicultural Politics. *Perspectives on Politics* (9), 606–609.
PRATT, G. (2005): Abandoned women and the spaces of the exception. *Antipode* 37 (5), 1052–1078.
SACCO, J. (1993): *Palestine.* Seattle.
SACCO, J. (2011): *Safe Area Gorazde: The Special Edition.* Seattle.
SAID, E. (2001) Homage to Joe Sacco. J. SACCO. *Palestine.* Seattle, i–v.
SAMSON, J. (2011): Le sujet de la guerre en bande dessinée. V. ALARY and B. MITAINE (Eds.): *Lignes de front: bande dessinée et totalitarisme.* Chêne-Bourg, 255–268.
SHARP, J. (2007): Geography and gender: finding feminist political geographies. *Progress in Human Geography* 31 (3), 381–387.
SHARP, J. (2011): Subaltern geopolitics: introduction. *Geoforum* 42 (3), 271–273.
TOAL, G. (1996): An anti-geopolitical eye: Maggie O'Kane in Bosnia 1992–93. *Gender, Place and Culture* 3 (2), 171–185.
WALKER, I. (1995) Desert stories or faith in facts. M. LISTER (Ed.): *The Photographic Image in Digital Culture.* London, 239–240.
WHITLOCK, G. (2006): Autographics: the seeing 'I' of the comics. *Modern Fiction Studies* 52 (4), 965–979.
WHITLOCK, G. and A. POLETTI (2008): Self-Regarding Art. *Biography* 31 (1), v–xxiii.

Chapter 6 | *Edward C. Holland*

POST-MODERN WITNESS: JOURNALISM AND REPRESENTATION IN JOE SACCO'S *CHRISTMAS WITH KARADZIC*

The comics journalist Joe SACCO is recognized and acclaimed for his work in Palestine, the former Yugoslavia, and other post-conflict zones. His comics are unique, simultaneously drawing from the tradition of underground and autobiographical comics – including the work of Robert Crumb and Art Spiegelman – and on-the-ground reporting by correspondents working in conflict zones. SACCO moves beyond the conventional ambit of journalistic work, using comics as a medium for persuasively communicating how the messy realities of geopolitics play out in their localized settings.

SACCO's work has gained increased attention from scholars in both comics studies and geography in recent years (WALKER 2010; CHUTE 2011; HOLLAND 2012; see also chapter 7). This work has considered a diverse set of topics, including the representation of trauma and violence to humanize conflict and the depiction of historical events through the narrative techniques made possible by comics' combination of image and text. In general, this existing work suggests that these narrative and artistic techniques can lead to more nuanced readings of the geopolitical events that SACCO portrays – a divergence from mainstream superhero comic books that employ the same format (HOLLAND 2012).

To extend this literature, in this chapter I incorporate another element of SACCO's work – his positionality as a journalist and how this role influences the narrative communicated in his comics – through a consideration of his short graphic narrative, *Christmas with Karadzic*. In this story, SACCO interrogates his role as a professional journalist, a self-reflective process that first reaffirms the practice of journalism, then unsettles it, and ultimately leads to SACCO's implicit questioning of who should be the central subject in the coverage of conflict.

To develop this argument, the chapter opens with a brief background on the study of graphic narrative. I then transition to discuss three topics of relevance to the work considered: a brief biography of Radovan Karadžić, the former Bosnian Serb political leader at the center of SACCO's comic; SACCO's own experience in Bosnia; and a synopsis of *Christmas with Karadzic*. Building on these subjects, I turn to a more focused discussion of SACCO's examination of his role as a journalist working in Bosnia at the conclusion of the war. SACCO's interrogation of this role – in particular his emotional reaction to encountering Karadžić – lends further credence to the argument that non-fiction graphic narratives disrupt and unsettle mainstream interpretations of conflict and other geopolitical events.

GRAPHIC NARRATIVE IN COMICS STUDIES AND GEOGRAPHY

Joe SACCO is one of a number of authors – Art Spiegelman, Marjane Satrapi, and Alison Bechdel are others – working in the graphic narrative format. SACCO has written and drawn graphic narratives on a variety of subjects in a range geographic locales: the consequences of the ongoing Israeli-Palestinian conflict for those living in the West Bank and Gaza; the difficulties facing internally displaced persons in Russia's North Caucasus; and recent work, in collaboration with the journalist Christopher HEDGES, which intersperses SACCO's visuals with critical reporting on capitalism and the unfettered marketplace in post-industrial America (HEDGES and SACCO 2012).

In a notable intervention on the format, Hillary CHUTE and Marianne DEKOVEN (2006, 767) define graphic narrative as "narrative work in the medium of comics," a wide-ranging description that encompasses a variety of topics, yet is also a concerted attempt to move away from the commonly used but potentially misleading term "graphic novel." Other authors, including WOLK (2007), employ the term "graphic narrative" in a similar fashion, to refer to any work that employs the combination of image and text that characterizes the comic book medium. The content of graphic narrative remains one of its most important elements; though the autobiographical is "arguably the dominant mode of current graphic narrative" (CHUTE 2008, 456), and the term is used here in reference to SACCO's non-fictional accounts, graphic narrative should be viewed inclusively with respect to format, content, and length.

In geography, the literature on popular geopolitics has primarily focused on the manner in which visual representations reproduce accepted and dominant geopolitical scripts. In its engagement with comics and graphic narrative, this work has considered the ways in which comics construct narratives of identity and belonging – particularly with the national community – through, for example, the representation of landscape (DITTMER 2005). Recent interventions have suggested that work on popular geopolitics move beyond this focus on representation and engage further with the ideas of audience response (DITTMER and DODDS 2008) and affect and emotion (see DITTMER and GRAY 2010), among other topics. In another attempt to extend this engagement, HOLLAND (2012) has suggested that graphic narratives can serve as a venue for questioning the prevailing narratives of popular media. As one of a suite of oppositional formats (including also political cartoons and satirical newspapers), graphic narrative relies on the blending of image and text, along with a varied set of narrative techniques, to craft an alternate interpretation of geopolitical scripts – what Ó TUATHAIL (1996) refers to as "an anti-geopolitical eye." In the critical geopolitics literature, the grounded perspective that graphic narrative employs can be contrasted with Cartesian perspectivalism, a worldview that distinguishes the observer from the observed and leads to untenable claims of objectivity and neutrality. This grounded perspective as a key element of graphic narrative is endorsed by scholars in other disciplines; to quote CHUTE (2008, 459): "The most important graphic narratives explore the conflicted boundaries of what can be said and what can be shown at the intersection of collective histories and life stories."

Further uniting the existing work in both comics studies and geography is a focus on the role of format and art in graphic narrative, including the spatiality of the comics' layout and the use of shifting temporalities. DAVIS (2005, 267) summarizes why graphic narrative should be distinguished from other formats, which are either solely visual or solely verbal: "the potential of the graphic narrative as a highly dynamic text, as opposed to the more static single-image narrative painting or plain text, determines the dialectic between text and image, providing creators with a wider range of artistic and imaginative possibilities." Moreover, the comics' format allows for communicative flexibility both temporally – for example, historical events can be depicted visually – and spatially – with frames and gutters producing a contingent and emergent reading of the text (DITTMER 2010). CHUTE and DEKOVEN (2006, 769) also acknowledge this melding of the spatial and the temporal in graphic narrative that has been theorized in geography (DITTMER 2010); "the form's fundamental syntactical operation is the representation of time as space on the page." In turn, one of graphic narrative's particular strengths is the incorporation of history and contingency into the story, a technique achieved through the interplay of structure and art.

The format also allows for flexibility with respect to representation and content; certain events are highlighted through the course of the narrative or particular features are emphasized in the depictions of characters. A more concrete example of how the use of representation and content complement one another from SACCO's work – though it is also employed in other well-known graphic narratives – is the insertion of the author into the narrative. SACCO's self-depiction is "a bit rubbery and cartoony" (GILSON 2005, n.p.); though he has refined his artistic style to be more representational and realistic – in comparison to his early work in *Palestine*, for example – his character is still rendered in the comic-book style and plays a central role in his graphic narratives. SACCO's self-depiction allows the reader to place themselves in the narrative; with respect to his cartoony representation, SACCO states that "people tell me the good thing about it is that they can step into the character because it is more of a formless character" (GROTH 2002, 61).

Taking this prior work as a point of departure, in this chapter I attempt to move beyond this discussion of how artistic depiction and the comics format shape the reader's experience and interpretation of graphic narrative. I suggest that scholars interested in comics and graphic narrative, regardless of disciplinary background, ought to attend to how autobiographical content – in this case, SACCO's professional identity as a journalist – affects the central ideas communicated through graphic narrative (see also DAVIS 2005). Foundationally, this reflection on the professional role of journalist is a content choice that allows SACCO to further situate himself in the historical and geographic contexts of the events depicted. In the subsequent sections, I explore these ideas as mediated through SACCO's work as a journalist in Bosnia, as detailed in his short graphic narrative *Christmas with Karadzic*.

RADOVAN KARADŽIĆ AND THE WAR IN BOSNIA

Before turning to this work, I first provide background on Radovan Karadžić and his role in the war in Bosnia in the mid-1990s. During the conflict, Karadžić was the political leader of the Bosnian Serbs. A founding member of the Serb Democratic Party (SDS) when it was established in July 1990, Karadžić was the key figure in the radicalization of the Bosnian Serbs' political platform in response to the breakup of Yugoslavia at the beginning of the 1990s. Following a referendum on independence held in early 1992, which was overwhelmingly approved by those who took part in the vote, Bosnia officially separated from Yugoslavia on 27 March 1992. The Bosnian Serbs, who had previously proclaimed a "Serb Republic of Bosnia-Herzegovina," responded with force on 1 April 1992, when Serb paramilitaries attacked the town of Bijeljina in northeast Bosnia (TOAL and DAHLMAN 2011). This marked the beginning of the Bosnian war.

Throughout the course of the conflict, Karadžić's role was primarily as political ideologue, articulating a vision of Bosnia where the country's three main religious-national groups – the Serbs, the Croats, and the Bosnian Muslims – were separated into distinct areas of territorial settlement. In May 1992, Karadžić explicitly laid out the Bosnian Serb position before a meeting of the Republika Srpska's secessionist parliament in Pale, a town located just to the east of Sarajevo, which served as the administrative center of the breakaway republic during the course of the war. This platform included the separation of national communities, to be tacitly supported through policies of ethnic cleansing, the outlining of the future borders of a Serbian entity within – or carved out from – Bosnia, and the Serb occupation of at least part of Sarajevo, the country's capital, to allow for the division of the city into similar ethnic enclaves (TOAL and DAHLMAN 2011). Throughout the war, Serb forces maintained a stranglehold on the city; sharpshooters and artillery targeted citizens from the heights around the capital. In a grotesque endorsement of Karadžić's vision, in early February 1994 Serb forces shelled the Markale marketplace in the historic center of the city, killing 68 and wounding more than 100. The July 1995 killing of approximately 8,000 men and boys in the UN-designated safe area of Srebrenica further confirmed that the ethnically motivated targeting of Bosnian Muslims was a central component of Karadžić's political platform.

After Srebrenica and further attacks by Serb forces against civilians in Sarajevo, in late August 1995 NATO launched an air campaign against Serb positions in tandem with a renewed offensive from Croat and Muslim military forces on the ground. In turn, the United States brokered a ceasefire, implemented in October 1995, which was followed by negotiations on a peace treaty held the next month. During these negotiations at Wright-Patterson Air Force base outside of Dayton, Ohio, Karadžić was left on the sidelines. Slobodan Milošević, President of Serbia proper, represented the SDS position during the crafting of the agreement (SELL 1999; TOAL and DAHLMAN, 2011). The General Framework Agreement, which detailed the obligations of the signatory parties at Dayton, was supplemented by a set of annexes that outlined provisions for a new constitution, the territorial division of Bosnia along the Inter-Entity Boundary Line, and the right of return for displaced persons, among other issues of dispute (CRAMPTON 1996; TOAL and DAHLMAN 2011).[1] The agreement implemented an almost equal (49/51) division of Bosnia's territory between the Republika Srpska (RS) and the Muslim-Croat federation, confirming the creation of a distinct Serbian territory through the policies of ethnic cleansing carried out during the previous three years.

The international community's military response to the Bosnia conflict prior to the events at Srebrenica was slow and ineffectual; it included a limited intervention by United Nations forces and the continuance of the arms embargo initiated in 1991 against all of the war's participants. The International Criminal Tribunal for the former Yugoslavia (ICTY) was established in May 1993 as a juridical mechanism for holding accountable those who committed war crimes during the course of the conflict (see JEFFREY 2009). The ICTY first issued an indictment against Karadžić in July 1995. He was charged with a range of crimes, including the destruction of property and sacred sites, the incarceration of Bosnian Croats and Bosnian Muslims in concentration camps, and the use of UN troops as human shields. In the wake of Srebrenica, subsequent charges by the tribunal – issued in November 1995 – further laid out the case against Karadžić for genocide, crimes against humanity, and violation of the laws or customs of war (see ROHDE 1998).[2] This was where the case against Karadžić stood in early January 1996, when Joe SACCO and his colleagues traveled to Pale in the hopes of interviewing the Bosnian Serb leader.[3]

JOE SACCO'S WORK IN BOSNIA

Joe SACCO first arrived in Sarajevo the previous fall, almost immediately after finishing the comic series that was eventually collected in *Palestine* (GROTH 2002; SACCO 2002). As SACCO states in his interview with Gary GROTH (2002, 59): "I

---

1 The text of the Dayton Accords is available at: http://www.state.gov/www/regions/eur/bosnia/bosagree.html

2 Charges of genocide, crimes against humanity, and the violation of the laws or customs of war were all included in the initial indictment, issued in July 1995. The text of the initial indictment is available here: http://www.icty.org/x/cases/karadzic/ind/en/kar-ii950724e.pdf; the text of the indictment detailing the crimes at Srebrenica is available at: http://www.icty.org/x/cases/karadzic/ind/en/kar-ii951116e.pdf. The indictment against Karadžić has been amended three times: in 2000, in 2008, and in 2009 (JEFFREY 2011).

3 Karadžić was arrested as he disembarked a tram in the Serbian capital of Belgrade on 21 July 2008 and was subsequently transferred to the ICTY. His trial at The Hague, which began soon after his arrest, is ongoing; a recent decision by the court to drop one of the charges of genocide against Karadžić gained international media attention, though 10 other charges – including one for genocide based on the actions of Serb forces at Srebrenica – remain in effect. For further background on this decision, see: http://www.guardian.co.uk/world/2012/jun/28/radovan-karadzic-cleared-genocide-charge.

went there [Bosnia] with some confidence because I felt that I had managed to achieve something journalistic" in that previous work. At the time, the Bosnian war was winding down. In Sarajevo, SACCO found that people were reticent to talk; "People [in the capital] were fed up with talking to journalists...They were jaded" (GROTH 2002, 60). Rather, SACCO was quickly drawn to the town of Goražde, which during the war was one of six Bosnian Muslim enclaves located within Serb-held territory in the eastern part of the country. One of the provisions of the negotiated ceasefire of October 1995 was access to Goražde for UN supply convoys; this land corridor was formalized in Annex Two of the Dayton Accords. SACCO traveled to the town with the initial convoys and was captivated by Goražde and the stories of its residents. As an outsider in an area that had been cut off from Sarajevo and the west for the past three-and-a-half years, he was welcomed by the town's residents: "When I got to Goražde...I'd be walking down the street and old women would be coming out of their homes saying, 'Come in, come in. Let me tell you about what happened here'" (GROTH 2002, 60). The resulting book, *Safe Area Goražde*, is SACCO's (2000) best-known work on the war in Bosnia.

In writings about his work, SACCO is consistently described as a comic journalist. Trained as a journalist, and acknowledging the autobiographical tradition in graphic narrative that first gained prominence during the 1980s, SACCO's graphic narratives are syncretic products that combine the elements of comics, journalism, and autobiography (GROTH 2002). SACCO endorses the comics format but also acknowledges that works that blend the format with reporting need to be done well: "More editors seem to be aware of how comics can be melded with journalism and are interested in experimenting with the form...but the work has to be good" (GILSON 2005). SACCO views the visual element as particularly valuable, as thought-provoking visuals are something to which people respond: "With comics you can put interesting and solid information in a format that's pretty palatable" (GILSON 2005). The late cultural critic Edward SAID (2002, ii–iii) makes a similar point in the introduction to SACCO's graphic narrative, *Palestine*:

> In a media saturated world in which a huge preponderance of the world's news images are controlled and diffused by a handful of men sitting in places like London and New York, a stream of comic book images and words, assertively etched, at times grotesquely emphatic and distended to match the extreme situations they depict, provide a remarkable antidote.

Taken as a whole, SACCO's *oeuvre* reflects his position as an artist who was also informed by the profession of journalism, with its on-the-ground elements of witness and reportage; in referring to *Safe Area Goražde*, he suggests that "the idea was to convey something that, at its heart, was journalistic. But I wanted to convey it in a way that would bring out the humanity of the people there" (GROTH 2002, 72). The artistic elements are intended to elicit an emotional response from his readers – as SACCO states in the same interview, "you want people to be moved by it" (GROTH 2002, 72). More importantly, the artistic renderings in his graphic narrative humanize the characters that are depicted. The intention is that readers care about the people that SACCO draws, and through this humanization the graphic narrative can approach the difficult themes that are at the heart of the story. In making a similar point elsewhere, SACCO has questioned the role of the media in the Bosnian conflict: "I don't blame the media for perpetuating this conflict, but in failing to fully inform the electorate, they prevent people from bringing their influence to bear on these situations, because they don't know what the hell is going on with them to begin with" (quoted in MCKENNA 2004, n.p.).

To distinguish this work from more mainstream journalistic outlets, comics journalism as practiced by SACCO – those "assertively etched, at times grotesquely emphatic and distended" images and words – requires the reproduction of visual details. When working in Bosnia, SACCO asked visual questions and took numerous photographs – for example, to ensure the accurate depiction of the first battle scene in *Safe Area Goražde* (SACCO 2000; GROTH 2002). SACCO needed to follow up with some of his subjects on the details of events to ensure this accuracy – "when you're doing an interview you can't anticipate everything you'll be drawing later" – even when these were difficult, even traumatic, themes (GROTH 2002, 63). The realism of SACCO's comics journalism is enhanced by this commitment to accuracy in representation, and furthers the format's communicative power. The visual precedes written explanations in conveying a range of real events from the conflict: the killing of children, the slow starvation of Goražde's elderly, and the difficulty of hiking overland through enemy lines. Like other non-fiction graphic narratives, SACCO's work relies on his personal experiences. This is not, however, a memoir; rather, as a journalist SACCO balances the narrative between himself and the people he meets and comes to know, their stories and how the war in Bosnia affected their lives. In the subsequent sections I further detail how SACCO's work is influenced by his professional role as journalist, a theme that is underscored in his short graphic narrative, *Christmas with Karadzic*.

A SYNOPSIS OF *CHRISTMAS WITH KARADZIC*

In addition to *Safe Area Goražde*, *Christmas with Karadzic* is one of three other works that emerged from SACCO's time in Bosnia.[4] The story first appeared in issue 15 of *Zero Zero*, an alternative comics anthology published by Fantagraphic Books, and was then recollected in 2005 along with another of SACCO's short profiles, "Soba," in *War's End: Profiles from Bosnia 1995–96*. At the time of republication, SACCO dedicated the story to the people at the ICTY and "all those who are working to arrest Radovan Karadzic" (SACCO 2005, 71).[5] However, this dedi-

4 In addition to *Soba*, which is collected together with *Christmas with Karadzic* and recounts the experiences of the eponymous main character in defending Sarajevo during the 1992-1995 war, *The Fixer* tells the story of Neven, SACCO's in-country handler when he returned to Bosnia in 2001.

5 SACCO (2012) recently published his reflections on the proceedings of the ICTY when it first opened in 1998 in *Foreign Policy* magazine. Neither KARADŽIĆ nor Ratko Mladić – the general who headed the Bosnian Serb forces during the war – had yet been arrested (see note

cation belies the main theme of the story: SACCO's journalistic work in postwar Bosnia. In referring to *Safe Area Goražde*, SACCO observes that "in Goražde I was around very compelling people that I knew would be the main focus of the book…they're more interesting [than me]" (GROTH 2002, 72). In comparison to his other works on Bosnia, in *Christmas with Karadzic* SACCO's role as a journalist takes center stage. Though other characters – including the two other journalists Kasey and Jack, as well as Karadžić and SACCO's landlady Rada – are important to the story, the main focus of the narrative is on SACCO himself.

The graphic narrative opens with SACCO and his colleagues speeding through post-Dayton Bosnia – and the checkpoints put up by local Bosnian troops – on their way to the town of Pale. The three are headed to the town in the hope of interviewing Karadžić. The night before, they received a tip that he would be attending religious services marking Orthodox Christmas at the town's church the next morning, January 7. A meeting with the Minister of Information serves as an opportunity for SACCO to provide some important elements of the backstory to the Republika Srpska and Karadžić himself: the lack of media access to Republika Srpska during the war (a policy that the journalists attribute to Sonja Karadžić, Radovan's daughter), Karadžić's *persona non grata* status at the end of the war in the eyes of the international community, and the policies of ethnic cleansing and the targeting of Sarajevo's multicultural landscape that Karadžić openly endorsed.

Arriving early the next morning, the trio is initially unable to locate the Bosnian Serb leader; to their surprise, the Orthodox Church is unguarded, with little indication that a service will soon be held. One of SACCO's colleagues, Kasey, grows frustrated with what seems like a false lead; to salvage the trip, he tries to record a spurt of gunfire and the peal of church bells to serve as background for an alternate story. The group's luck improves when they receive confirmation that Karadžić will, in fact, attend the 9am service at the town's church. Prior to this service Kasey conducts a six-minute interview with Karadžić. During the interview, SACCO (2005, 59) characterizes Karadžić as "generous" and "dignified," and the whole scene as something not beyond the ordinary for the crowd of onlookers. Through the course of the service, Karadžić listens respectfully. SACCO, meanwhile, meditates on the atrocities committed under Karadžić's leadership and one of his most notorious statements: "Sarajevans will not be counting the dead. They will be counting the living." The narration of the scene communicates SACCO's (2005, 59) divergent reaction to seeing Karadžić in person: "I feel nothing intimidating about his presence, nothing extraordinary about this man indicted by the International War Crimes Tribunal for crimes against humanity, a man I have despised with all my heart for years…" The service progresses, though SACCO and his colleagues leave roughly half an hour through; Kasey celebrates his coup and dreams about selling the story as an exclusive to American media outlets.

---

three, above); when SACCO attended the tribunal, only 30 criminals stood trial. Yet he had hope for the ICTY; despite its failure at its opening to have apprehended the "big fish," SACCO suggests that "justice is worth pursuing for its own sake" (SACCO 2012, 6).

The graphic narrative concludes on a more prosaic note, with SACCO returning to Sarajevo to share an Orthodox Christmas meal with his landlady Rada – who, like Karadžić, is Serbian – and her family. A panel in the narrative depicts the view from Rada's apartment; through a taped-up window, SACCO identifies the Serb-controlled area of Grbavica and the Holiday Inn hotel, which served as the home base of the international press corps that came to Sarajevo during the war. The walk through Sarajevo to Rada's sister's apartment gives SACCO an opportunity to incorporate a series of background images that underscore the post-conflict condition of the city: a tank (with the markings of IFOR, the NATO-led peacekeeping force) patrolling the streets, a shelled home, and a burned-out car. After eating, the group tunes in to the Bosnian Serb newscast from Pale, which reports on Karadžić's attendance at the morning service. SACCO can see himself in the background of the television shot, though he had tried to duck out of view when the camera swung towards him. SACCO summarizes his feelings about the encounter in closing the narrative: "I'd felt nothing more in his presence, nothing, not revulsion, not loathing, no matter how hard I'd tried...In fact, going to see him was the most fun I'd had at Christmas in years" (SACCO 2005, 65; emphasis in the original).

POST-MODERN WITNESS: JOE SACCO'S POSITIONALITY AS JOURNALIST

Christmas with Karadzic effectively conveys the consequences of the war in Bosnia: the tenuous nature of the post-Dayton peace; the marginalized position of Karadžić and those Bosnian Serbs who had supported his political platform; and the destruction of Sarajevo after a three-year-long siege. Beyond the narrative's content, in this section I consider the role of journalism, and SACCO as journalist, in first reaffirming and then unsettling interpretations of the profession and its place in conflict zones. Complementing the storytelling and artistic techniques made possible by the graphic narrative format, this interrogation of the journalistic role ultimately leads to a questioning of who should serve as subject in such work.

To reiterate, the interpretation of SACCO's *Christmas with Karadzic* should necessarily be situated within wider debates in the disciplines of geography and comics studies. Comic books have been acknowledged by other authors to be important sites for the communication of ideological positions (MCALLISTER et al. 2001); DITTMER (2005, 641), for example, refers to the ideological nature of *Captain America* comics in their attempt to make "cultural claims to geopolitical truth." Recognizing the inadequacy of the mainstream media in Bosnia and other places, SACCO's graphic narrative does not present such parsimonious positions in its content; "the graphic narrative's complicated mix of subjective narration and objective history, image and text, personal and collective when dealing with historical trauma causes it to function as a subjective documentary" (NAYAR 2009, 152). To reiterate a point made previously, this grounded perspective contrasts and challenges the Cartesian perspectivalism of mainstream geopolitics. Furthermore, the personalization of experience as facilitated by SACCO's position as journalist results in the privileging of certain aspects of memory, emotion, and interpretation in the narrative.

In its first half, *Christmas with Karadzic* juxtaposes two elements of the journalistic enterprise: the thrill of chasing the story and the more mundane aspects of waiting – and hoping – for that story to materialize. The three journalists come to Pale on a tip; their meeting with the Minster of Information is encouraging, but does not conclusively confirm that Karadžić will be at the Christmas morning service. When they return to Pale the next day, the trail seems to have gone cold; to quote Kasey: "We're just being very unlucky this morning, that's what happening" (SACCO 2005, 56). This turn in luck is depicted as frustration, which further accentuates the tenuousness of securing the story. At the same time, this chase is positioned against a background of the journalists' uncertain status in the Republika Srpska. After first meeting with the Minister of Information, the three are then called to Pale's International Press Center to meet with Sonja Karadžić (who served as the Center's Head during the war), in an effort to guarantee the legality of their status in the territory; unsuccessful in gaining the proper permits – the building housing the Press center appears to be empty – the three remain concerned that they will be locked up and their car confiscated. This negotiation of the field parallels the practical and strategic questions that academics must wrestle with when conducting research, including the legality of their presence in a particular place, the potentially hostile reaction of local authorities to their work, and the possibility of interrogation by these authorities (KATZ 1994).

The graphic narrative also includes a number of scenes that suggest that journalism and its practice can be routine and even banal. After finding out that Karadžić will, in fact, be attending the Christmas morning service, the group sits down for coffee and cigarettes, an everyday moment that gives the three journalists a chance to speculate on how the interview will go. This everyday scene also offers an opportunity to discuss another running theme throughout the story: the attractiveness of Jugoslava, who works for Republika Srpska Television and who provided the tip that Karadžić would be in Pale on Christmas morning. The image of three male reporters sitting around discussing the hyper-sexualized attributes of one of the female figures in the story is hackneyed, though arguably intentionally so – Jugoslava is more of a representation than a character. Throughout the course of the story, her face is never shown – instead, she is depicted from the chest and waist down. As SACCO (2005, 47) describes the scene: "we were praying for more stairs, for more miniskirt and bottoms, but we got only two flights' worth before she ushered us into the office" of the Minister of Information. This thread and the discussions that SACCO reproduces mirror the gendered history of journalism identified by FARISH (2001, 276, emphasis in the original); the journalist is "a deeply masculine figure" who produces an "authoritative *geography*" from what he witnessed. During the car trip back to Sarajevo on Orthodox Christmas Eve, Kasey, in responding to the group's rule that "you don't make it with a Chetnik," waxes philosophical and geographic: "you take her out of here, you take her to California – she wouldn't be a Chet" (see also Figure 6.1; SACCO 2005, 52).[6]

The arrival of Karadžić marks a point of rupture in the graphic narrative. SACCO had previously focused on his and his colleagues' efforts to secure an interview with Karadžić – the more prosaic, at times even banal, elements of journalism. From this point on the story's main theme is SACCO's personal experience and reflection. Despite the buildup to Karadžić's appearance as occurs throughout the first half of the graphic narrative, the moment itself is depicted in an understated fashion. SACCO and his colleagues stand with a larger group in the churchyard when Kasey spots a Mercedes; the first scene depicting Karadžić on Christmas morning has him in the background, surrounded by bodyguards and accompanied by his wife. In the set of images that illustrate Kasey's interview with Karadžić, SACCO is in the background of only one – his face is unreadable, with any emotional expression masked in part by the glasses that cover his eyes. This ambivalent self-representation foreshadows SACCO's feelings towards Karadžić as expressed during the Christmas morning church service.

Over the next page and a half, SACCO reflects on Karadžić's actions as leader of the Bosnian Serbs; "During the service, I keep looking over at him waiting for something to sink in, but it never does…not the rapes, not the concentration camps, not the 'cleansing,' not the throats slit and the bodies dropped into the Drina" (SACCO 2005, 59–60). SACCO then tries to find a specific point of focus, to concentrate his feelings towards Karadžić. He settles on Karadžić's words foretelling the fate of Sarajevans; even this, however, does not produce the response that SACCO anticipated. Throughout the scene in the church, moreover, SACCO himself is not depicted, and does not appear again until he leaves the service with his colleagues, celebrating the coup of conducting the interview. Instead, the reader relies on his internal monologue to understand his reaction. And the reader is left with an impression that SACCO's emotional reaction was not what he anticipated in light of the war crimes that Karadžić has committed – rather, it was ambivalent and confused. SACCO has elsewhere acknowledged this complicated response to seeing Karadžić in person. In an interview with *Mother Jones* magazine (GILSON 2005, n.p.), he describes the helplessness he felt during the church service thus:

> I'd been stewing about Karadžić for a long time. Sometimes when you're a journalist you put aside your personal feelings about someone because it's more important to get the story. Then when you finally meet them, you're not prepared. I realized I had nothing to say to this man. What could I say? "How could you do this?"

6 The term "Chetnik" is used here to refer to Serbian nationalists. Its origins lie in Serbian resistance to the Ottoman Empire prior to World War I; the Chetniks were also part of the anti-Nazi resistance in Yugoslavia during World War II. The term was commonly employed in Bosnia to refer to those individuals who supported Karadžić's political platform as described previously.

SACCO continues: "That story [*Christmas with Karadzic*] is really about the fight between being a journalist and being someone who actually *cares* about what they write about" (GILSON 2005, n.p., emphasis added).

In the graphic narrative, SACCO anticipates a more personal emotional reaction than seeing Karadžić actually elicits. As such, the interpretation of SACCO's encounter with Karadžić should be informed by the role of emotion – an emergent theme in the critical interrogation of geopolitics that considers the production and circulation of sentiments of fear, anger, and hope (HENDERSON 2008; PAIN 2009). DALBY's (2010, 283) recent prospectus for the future of critical geopolitics expresses uncertainty over the link between emotional geographies and "the discourses used to legitimate the practices of violence." In other words, this suggests questioning the link between emotion and the broader project of critical geopolitics, given its traditional emphasis on the interrogation of geopolitical scripts and narratives. PAIN (2009, 478) offers what can be read as an answer to this question – "we might see emotions not just as blank canvasses…but as situated, historicized and relational – already formed and always changing – and affecting politics, as much as they are affected by politics, at a range of scales."

As he himself acknowledges in the quote provided above, SACCO's emotional reaction must be situated with respect to his professional affiliation as a journalist. A feeling of disgust and a desire to confront Karadžić characterizes the narrative up to the meeting with the Bosnian Serb leader, substantiated through the brief history of Karadžić's actions and statements during the war. Importantly, at the moment of the encounter in the text, I argue that what occurs in *Christmas with Karadzic* is a reversal of the journalistic role. The profession demands objectivity and distance in reporting of the story, regardless of the emotions that underlie this scripted ambivalence. In his graphic narrative, SACCO openly acknowledges his emotional position – he despises Karadžić and condemns his actions during the Bosnian war. SACCO's story as communicated in *Christmas with Karadzic* is far from ambivalent – his professional product is clearly positioned as a condemnation of Karadžić's actions. His personal, emotional engagement with Karadžić is, however, much more complex, as his actual reaction departs from the expectation of hatred and anger. While journalists often mask their emotional reaction in the interest of objectivity, SACCO's narrative presents a surprising and confusing sentiment – a lack of the anticipated emotional reaction. In turn, there is no need for false objectivity in the process of reporting the story.

In turn, what is questioned through SACCO's work is the detached, disembodied Cartesian perspectivalism commonly associated with the journalistic profession. Following PAIN (2009), SACCO's emotions are historicized and relational, yet also unanticipated and complex, and communicated through the graphic narrative format. The journalistic frames of the Bosnian war most commonly reproduce the broader geopolitical scripts of "quagmire" and "Holocaust"; these scripts also conditioned the apathetic political response of Western leaders to the conflict (see Ó TUATHAIL 1996). This emerges, in the mainstream media, as a performance of the journalistic role. Such performativity – the production of the subject through reiterated acts that results in a form of uncritical acceptance – entrenches certain readings of geopolitical space through iteration and reproduction; in discussing performativity, political geographers "are dealing with an assemblage of practices…[including] the representational technologies of popular geopolitics – which together produce the effect they name, stabilizing over time to produce a series of spatial formations through the performance of security" (BIALASIEWICZ et al. 2007). Journalists, by masking their true emotional reaction in the interest of objectivity, play a role in the reproduction of these dominant scripts and frames.

Rather than reproducing the expectations of journalism, SACCO's appeal is to question such expectations, through a self-reflective interrogation of the profession that emphasizes his personal – and conflicted – emotions, and enhanced by the artistic rendering of geopolitically complex events made possible by the graphic narrative format. While SACCO unambiguously communicates the feeling he anticipated, his actual reaction is distinct from this expectation. The graphic narrative, in turn, is rooted in the complication of positionalities, as artist and journalist. SACCO's position within his profession is complex; he does not simply accept the frames constructed around the Bosnian war or its central villains. As protagonist in the story, SACCO is located in the profession of journalism as a "discursive mode," an ontology that structures certain conventions and expectations (BIALASIEWICZ et al. 2007). In opening the narrative, SACCO makes clear his position towards Karadžić – it is one of unambiguous disgust. The rupture in these feelings happens during the encounter, and SACCO finds that he does not need to feign objectivity in communicating the experience; rather, this lack of emotional response in complex and unanticipated, yet fully explored through the graphic narrative format.

This unsettling of the journalist's role is consolidated in the final pages of the graphic narrative, where SACCO implicitly questions who and what should be the subject of journalism in conflict zones. Rada, SACCO's Serbian landlady, offers a stark contrast to Karadžić (see Figure 6.2); she remained loyal to Bosnia as a multiethnic state and continued to live in Sarajevo during the course of the war, despite being targeted by snipers in her apartment. The apartment next to her sister's was destroyed by artillery fire. The realities of life in war-time Sarajevo have taken their toll on both Rada and the city itself; SACCO notes that more than 10,000 residents of the city were killed during the course of the war. In contrast, the family scenes on the final page of the narrative reflect sentiments of hope and survival through conflict.

In turn, this leads to the questioning of who should serve as the subject in any consideration of geopolitical topics. SACCO's implicit critique of the subject is reflected most clearly in feminist approaches to geopolitics. As DOWLER and SHARP (2001, 172) write in one of the foundational pieces of this literature, it is "in the processes of subject creation – and resistance to this – that we can see how politics works." Rada's character implicitly challenges the stereotypes that revolved around Serbs during the Bosnia conflict. Moreover, SACCO's focus in the narrative's final few pages shifts away from Karadžić as political leader of a desired state to a finer scale of analysis, that of the home (HYNDMAN 2001). Karadžić is the central subject of mainstream journalistic accounts; he is, after all, who Kasey wants to interview in order to sell his story in the Western media. By inserting Rada into the narrative, SACCO is questioning the value of journalistic accounts that singularly focus on political leaders; the reality is more nuanced and complicated than these frames convey.

In sum, SACCO's engagement with Karadžić must be read considering two points: 1) the situated and relational nature of emotions with respect to the expectations of the profession of journalist and 2) the wider function of graphic narrative as autobiographical text that works to communicate the complex experience of the author to the readership. The lack of a strongly negative reaction by SACCO and the concluding admission of enjoying the day chasing Karadžić reflect SACCO's complicated emotional engagement with his subject and the practice of journalism more broadly. I suggest that this disconnect between expectation and reaction is attributable, at least in part, to SACCO's professional affiliation as a journalist – a positionality that is influenced by the journalistic frames constructed around the Bosnian war – and his ascription of guilt to the Bosnian Serb leader for the war crimes committed during the preceding three and a half years. What Ó TUATHAIL (1996, 175, emphasis in the original) refers to as "the modern practice of highbrow journalism with its organizing mythology of *objective* reporting by a naked eye" is contrasted with SACCO's attempt to communicate his emotional response, an approach that endorses the grounded perspective cultivated by other journalists working in Bosnia, specifically Maggie O'Kane as considered by Ó TUATHAIL (1996; see also HOLLAND 2012).

CONCLUSION: PROFESSION AND POSITIONALITY IN THE CRITICAL GEOPOLITICS OF GRAPHIC NARRATIVE

In this chapter, I have furthered the argument that graphic narrative provides a medium for representation and communication distinct from commonplace interpretations of geopolitical events. While SACCO's narrative and artistic techniques – including the insertion of himself as character into the narrative – complicate the reader's response to the material with which he engages, SACCO's work is not only representational. It also interrogates the professional role of the journalist in conflict zones. In *Christmas with Karadzic* SACCO emphasizes the professional half of the comics journalist label frequently ascribed to his work. The narration of the autobiographical as communicated through graphic narrative unsettles how the reader reacts to the singular geopolitical event; importantly for SACCO as author, it also serves as a medium for exploring the divergence of reactionary expectations and anticipated emotions. The graphic narrative depicts SACCO's complex emotional reaction while simultaneously using the "layers of meaning" that characterize the comic book medium to influence how the reader interprets his encounter with Karadžić (DAVIS 2005, 267).

Joe SACCO's work is central to understanding the distinction between journalistic framings of war and emotional responses to it as mediated through art. Other examples of graphic narrative reflect the need to consider the author's positionality in addition to the narrative and artistic techniques that characterize the comics format. Art Spiegelman's role as the son of two survivors of the Holocaust, Marjane Satrapi's experiences as a youth in revolutionary Iran and then in Europe as a *de facto* exile from her homeland, and Guy Delisle's work in North Korea as an animation cartoonist brought in to work on a film for the repressive regime all reflect a similar need. While some work in comics studies has pursued this question, scholars working in geography have not as of yet. The broader aim of this chapter, then, has been to acknowledge the subjectivity of the author in graphic narrative's ability to offer alternate interpretations of mainstream geopolitical events.

REFERENCES

BIALASIEWICZ, L., D. CAMPBELL, S. ELDEN, S. GRAHAM, A. JEFFREY and A. J. WILLIAMS (2007): Performing Security: The Imaginative Geographies of Current US Strategy. *Political Geography* 26 (4), 405–422.

CHUTE, H. (2008): Comics as Literature? Reading Graphic Narrative. *PMLA* 123 (2), 452–465.

CHUTE, H. (2011): Comics Form and Narrating Lives. *Profession 2011*, 107–117.

CHUTE, H. and M. DEKOVEN (2006): Introduction: Graphic Narrative. *MFS: Modern Fiction Studies* 52 (4), 767–782.

CRAMPTON, J. (1996): Bordering on Bosnia. *Geo Journal* 39 (4), 353–361.

DALBY, S. (2010): Recontextualising violence, power and nature: The next twenty years of critical geopolitics? *Political Geography* 29 (5), 280–288.

DAVIS, R. G. (2005): A Graphic Self: Comics as Autobiography in Marjane Satrapi's Persepolis. *Prose Studies* 27 (3), 264–279.

DEBRIX, F. (2008): *Tabloid Terror: War, Culture and Geopolitics*. New York.

DITTMER, J. (2005): Captain America's Empire: Reflections on Identity, Popular Culture, and Post-9/11 Geopolitics. *Annals of the Association of American Geographers* 95 (3), 626–643.

DITTMER, J. (2010): Comic Book Visualities: A Methodological Manifesto on Geography, Montage and Narration. *Transactions of the Institute of British Geographers* 35 (2), 222–236.

DITTMER, J. and K. DODDS (2008): Popular Geopolitics Past and Future: Fandom, Identities and Audiences. *Geopolitics* 13 (3), 437–457.

DITTMER, J. and N. GRAY (2010): Popular Geopolitics 2.0: Towards New Methodologies of the Everyday. *Geography Compass* 4 (11), 1664–1677.

DOWLER, L. and J. SHARP (2001): A Feminist Geopolitics? *Space and Polity* 5 (3), 165–176.

FARISH, M. (2001): Modern Witnesses: Foreign Correspondents, Geopolitical Vision, and the First World War. *Transactions of the Institute of British Geographers* 26 (3), 273–287.

GILSON, D. (2005): The Art of War: An Interview with Joe Sacco. *Mother Jones* July/August 2005. http://www.motherjones.com/media/2005/07/joe-Sacco-interview-art-war (accessed July 14, 2012).

GROTH, G. (2002): Joe Sacco, Frontline Journalist: Why Sacco went to Goražde. *The Comics Journal* Special Edition Winter 2002 – Volume 1, 55–72.

HEDGES, C. and J. SACCO (2012): *Days of Destruction, Days of Revolt*. New York.

HENDERSON, V. L. (2008): Is there Hope for Anger? The Politics of Spatializing and (Re)Producing an Emotion. *Emotion, Space and Society* 1 (1), 28–37.

HOLLAND, E. C. (2012): "To Think and Imagine and See Differently": Popular Geopolitics, Graphic Narrative, and Joe Sacco's "Chechen War, Chechen Women." *Geopolitics* 17 (1), 105–129.

HYNDMAN, J. (2001): Towards a Feminist Geopolitics. *The Canadian Geographer/Le Géographe Canadien* 45 (2), 210–222.

JEFFREY, A. (2009): Justice Incomplete: Radovan Karadžić, the ICTY, and the Space of International Law. *Environment and Planning D: Society and Space* 27 (3), 387–402.

JEFFREY, A. (2011): Ratko Mladić: Haematological Narratives of Violence. *Society and Space Blog*. http://societyandspace.com/2011/05/27/haematological-mladic/ (accessed June 27, 2012).

KATZ, C. (1994): Playing the Field: Questions of Fieldwork in Geography. *The Professional Geographer* 46 (1), 67–72.

MCALLISTER, M. P., E. H. SEWELL, JR. and I. GORDON (2001): Introducing Comics and Ideology. M. P. MCALLISTER, E. H. SEWELL, JR. and I. GORDON (Eds.): *Comics & Ideology*. New York, 1–13.

MCKENNA, K. (2004): Brueghel in Bosnia. *LA Weekly*, 1 January 2004. http://www.laweekly.com/2004-01-01/news/brueghel-in-bosnia/ (accessed June 22, 2012).

NAYAR, P. K. (2009): Scar Cultures: Media, Spectacle, Suffering. *Journal of Creative Communications* 4 (3), 147–162.

Ó TUATHAIL, G. (1996): An Anti-Geopolitical Eye: Maggie O'Kane in Bosnia, 1992–1993. *Gender, Place and Culture* 3 (2), 171–185.

PAIN, R. (2009): Globalized Fear? Towards an Emotional Geopolitics. *Progress in Human Geography* 33 (4), 466–486.

ROHDE, D. (1998): *Endgame: The Betrayal and Fall of Srebrenica, Europe's Worst Massacre since World War II*. Boulder, CO.

SACCO, J. (2000): *Safe Area Goražde: The War in Eastern Bosnia 1992–1995*. Seattle, WA.

SACCO, J. (2002): *Palestine*. Seattle, Washington.

SACCO, J. (2005): Christmas with Karadzic. *War's End: Profiles from Bosnia 1995–96*. Montreal, QC, 45–65.

SACCO, J. (2012): Portrait of the Hague as a Young Court. *Foreign Policy*, 16 May 2012. http://www.foreignpolicy.com/articles/2012/05/16/portrait_of_the_hague_as_a_young_court?page=0,0 (accessed June 25, 2012).

SAID, E. (2002): Homage to Joe Sacco. *Palestine*. Seattle, Washington, i–v.

SELL, L. (1999): The Serb Flight from Sarajevo: Dayton's First Failure. *East European Politics and Societies* 14 (1), 179–202.

TOAL, G. and C. T. DAHLMAN (2011): *Bosnia Remade: Ethnic Cleansing and its Reversal*. New York.

WALKER, T. (2010): Graphic Wounds: The Comics Journalism of Joe Sacco. *Journeys* 11 (1), 69–88.

WOLK, D. (2007): *Reading Comics: How Graphic Novels Work and What They Mean*. Cambridge, MA.

PART 3 | SPACE AND COMICS THEORY

Chapter 7 | *Julia Round*

WE SHARE OUR MOTHER'S HEALTH: TEMPORALITY AND THE GOTHIC IN COMIC BOOK LANDSCAPES

The body of the urban subject in the nineteenth century was frequently read in terms of the body politic of the city, as can be seen in political and recreational literature of the time (STALLYBRASS and WHITE 1986). But given that our understanding of the body has evolved so much over the past two hundred years, how does this idea now apply to twentieth and twenty-first century narratives, and specifically the visual landscapes of comics? This chapter will use gothic theory to analyse the uses of spatiality in the landscapes of three contemporary comics and especially their links to temporality (given the time-as-space narratology of comics). After noting the (more traditional) strategies used in *From Hell* (Alan MOORE/Eddie CAMPBELL, 1988–89) and *The Invisibles* (Grant MORRISON/various, 1996–2002), the main discussion will focus on *The Walking Dead* (Robert KIRKMAN/Charlie ADLARD, 2003–present).[1] Although set in diverse cultures and eras, these comics seem peculiarly gothic in their treatments of time and space. *From Hell* straddles the boundary between fact and fiction, being a meticulously researched yet self-proclaimed 'melodrama' that portrays the Jack the Ripper killings in Victorian London as a Masonic conspiracy under Royal direction: violent acts which bend space and time to 'deliver' the twentieth century in all its brutal misogynistic glory. *The Invisibles* focuses on a battle between terrorist cells from the 'Invisible College' and the 'Outer Church' whose characters can use magic, time-travel, enter fictional worlds, change identity, and interrogate the notion of reality itself as language becomes object and our world is redefined as a hologram of two overlapping universes. In *The Walking Dead* a still and silent post-apocalyptic American landscape becomes the setting for a group of survivors' struggles in the wake of an (unexplained) zombie uprising. It will be argued that gothic tropes such as fragmentation and inversion inform the construction of all of these landscapes, and that these in turn reflect the contemporary construction of identity as posthuman, multiple and crypotomimetic. 'Mother nature' is scarred by cities and technology and, in this sense, we share her health as our identity is problematised. The article will conclude by reflecting on its observations in the context of the comics narratologies expressed by theorists such as Charles HATFIELD and Thierry GROENSTEEN.

---

1 Please see the author's articles 'London's Calling' (2008) and 'Gothic and the Graphic Novel' (2010) for a more detailed discussion of embodied cities in *From Hell, Neverwhere* and *The Invisibles*.

LONDON: A DIVIDED BODY/CITY

Peter STALLYBRASS and Allon WHITE consider Parliamentary documents, social reform texts, psychoanalytic reports, journals, and fiction from nineteenth-century London, arguing that this literature demonstrates a tendency to view the city as an 'othered' body: "As the bourgeoisie produced new forms of regulation and prohibition governing their own bodies, they wrote ever more loquaciously of the body of the Other – of the city's 'scum'" (1986, 126). They also note a distance in these descriptions, which survey their surroundings with reference to a social divide that, although strenuously insisted upon (and demarked by lifestyle, dress, speech and much more), was simultaneously transgressed on a daily basis as people of many classes shared the overcrowded streets.

They conclude that "The axis of the body is transcoded through the axis of the city, and whilst the bodily low is 'forgotten', the city's low becomes a site of obsessive preoccupation, a preoccupation which is itself immediately conceptualized in terms of discourses of the body" (1986, 145). We can see an equivalent to this in twentieth-century comics such as *From Hell* and *The Invisibles*, in which London's lights and leylines are described as "veins" (14.10).[2] *From Hell* in fact bases its plot around the class divide noted by STALLYBRASS and WHITE, as the story's events are first started by the marriage of a prince and a shopgirl. Here, the divide between rich and poor is emphasised and simultaneously transgressed, for example in chapter five which juxtaposes William Gull's daily routine with Polly Nicholls's search for a bed for the night. As Royal Surgeon-in-Ordinary, Gull's upperclass lifestyle is reinforced by the appearance of his panels, which are drawn in soft, gentle shading, while the experiences of the prostitute Polly are represented in the scratchy linework that characterises the rest of the comic. Eddie CAMPBELL's artistic style emphasises the class divide between the characters even as the panels place them alongside each other; and the style of the comic-as-a-whole absorbs them all into the same living city.

This idea of the city as a (living) body is also used in *The Invisibles*, and again tied to tropes of transgression and sex. For example landmarks such as the Millennium Dome and the Canary Wharf Tower are described by Lord Fanny as "tits and a dick" (2002, 75) and she names the city a "travesty." In Latin American terms (Fanny is a Brazilian transvestite witch) *travestis* have been described as a 'third gender' and the term generally refers to a biological male with a feminine gender identity – though, in contrast to transsexuality, they do not suffer from gender dysphoria and do not wish to become female. This phrasing therefore defines London as a city that transcends simple binaries such as biological sex and class – as Fanny's name also represents; since it combines upper-class masculinity ('Lord', relating to the English peer system) with lower-class femininity ('fanny' is British slang for female genitals). Despite Fanny's nationality, her moniker and

2 References from *From Hell* will be cited in this format, which refers to Chapter 14, Page 10.

comments thereby clearly relate to the comic's initial location of London/England.

All of these transgressive images and phrases recall the gothic's reliance on inversion. For example, the gothic vampire is argued by Fred BOTTING (1996, 149) to be "a pure inversion [...] the mirror and shadow of Victorian masculinity"; a perversion of the masculine. The vampire's transgressive sexuality is created through this doubled and inverted gendering (as they penetrate with a bite, yet were also penetrated at the time of their 'turn'). While vampiric physicality is generally non-sexual (they do not have sex), their demeanour has nonetheless been heavily sexualised in contemporary culture: like the *travesti* the biological and cultural do not align. The Gothic allows multiple possibilities for sexual and gendered identity, transgressing simple boundaries and binaries.

Organs such as skin and bowels are also used to metaphorically describe aspects of London in *The Invisibles* (1996, 59; 2002, 145). This imagery seems significant in terms of the gothic concept of abjection. As well as Mr Six's description of the Underground rail network as the city's "bowels", Jack Frost rants about the "blood and sweat and shit" that London is built upon (1996, 106). Julia KRISTEVA defines the abject as that which confronts us with the knowledge that our bodies are also sites of otherness, such as: "A wound with blood and pus, or the sickly, acrid smell of sweat, of decay [...] refuse and corpses *show me* what I permanently thrust aside in order to live. These body fluids, this defilement, this shit are what life withstands, hardly and with difficulty, on the part of death. There, I am at the border of my condition as a living being" (3).

These landscapes can therefore be defined as gothic settings, made up of abject images and a transgressive framework. As well as the class divide already noted, *From Hell* invokes a split between the visible and the unseen/occult, as argued by Lisa COPPIN, for example when we are shown the image of Prince Albert's ghost (12.1). I have argued elsewhere that this divide is apparent throughout the book, as many of its events can be defined according to a tension between the physical and the spectral – or, space and time. For example the Ripper murders are both excessively corporeal *things* (considering the amount of blood and violence) and magical *events* with ritual significance. The role of seeing is emphasised through aesthetic means and iteration: "Look at it Netley. Can you SEE? Can you SEE it, Netley? [...] Light, Netley. Did you see? She was full of light," (5.33 [caps as in text]). Similarly, Nicholas Hawksmoor's churches are physical structures yet imbued with bodily significance and occult consequence: "Though Wren forbade interment on Church premises, Hawksmoor declined; sank his foundations in the plague-pits here, seeking the nourishment of which the Druids spake...so that his will, his personality encoded into stone might thus endure throughout the centuries" (4.32). These symbols thus have equal significance in both space and time.

The Invisibles also relies upon a notion of a divided world: defining our universe as a hologram made up of two overlapping universes and arguing that this is what makes transgression and change possible. Lord Fanny says: "In your world there is only ever the ugly caterpillar, just as in the world above, only the beautiful butterfly exists. […] / Here, in my world, there is change. […] Only here," (2000, 166.1–2). Duality is what makes change possible – the combination of two states, which allows movement and transgression. And change, of course, requires time in order to occur; unlike stasis which can exist in space alone. The links between time and space are brought to the surface in these comics' treatments of their locations and overtly commented upon.

THE DIVIDED BODY AND URBAN PERFORMATIVITY

These locations are not just bodies, then, but divided bodies – a notion that becomes significant when considering the superhero. The superhero has always traditionally been a psychogeographic figure; inextricably linked to his/her landscape. Some of the industry's most famous examples evidence this: Superman resides in a sunlit Metropolis, while Batman lurks in Gotham after dark. Personality, ethos and appearance all align with the hero's environment. As Neil GAIMAN put it (in the voice of the Riddler): "When is a *man a city*? When it's Batman or when it's Gotham," (GAIMAN 1989, n.p.). And, of course, there are other heroes, such as Alan MOORE's revisionist Swamp Thing and Neil GAIMAN's Black Orchid, who are quite literally embodiments of their environments.

Comics writer and editor Denny O'Neil has famously commented that Gotham and Metropolis are two sides of the same city: "Gotham is Manhattan below Fourteenth Street at 3am, November 28 in a cold year. Metropolis is Manhattan between Fourteenth and One Hundred and Tenth Streets on the brightest, sunniest July day of the year" (quoted in BOICHEL 1991, 9). Other writers are in agreement: Frank Miller describes Metropolis as "New York in the daytime; Gotham City is New York at night" (quoted in MACDONALD and SANDERSON 2006, n.p.) and although there have been dissenting opinions from others, such as comics artist Neal Adams and film director Christopher Nolan who dissociated Gotham from Manhattan in favour of Chicago (URICCHIO 2010, 122), the image of a divided city prevails in the DC universe.

However, as our understanding and conception of the body become more complicated, then, so do the images we use to represent it. The twenty-first century body is multiple and posthuman. As Judith Butler argues, gender is performative and no longer tied to biological sex. Physicality is enhanced and (in many Western cultures) appearance is modified and taken as self-expression. This leads to a model of identity that is fragmented and multiple. Communication advances invite us to construct and perform our identity/ies in a multitude of spaces, both physical and virtual. We can create as many alter egos and avatars as we wish, and the construction of personality thereby becomes a further performance: posthuman, multiple and crypotomimetic.

William URICCHIO (2010, 129–130) argues that a comic-book city such as Gotham is best approached not as an objective location, but rather as "a site of performance and a set of enacted practices [...] not so much a site of contradiction as a palimpsest of half-remembered episodes, out-of-sequence encounters with the text, selectively accruing landmarks and ever-reinvented spaces." He argues that this type of reading accords with the modern conception of cartography, which has "shifted over the past few decades from a largely technical enterprise concerned with the production of objective representations, whether static or dynamic, to approaches more behavioralist and culturalist in character" that acknowledge "marginal social formations" and the "transient and dynamic character" of social spaces.

URICCHIO (2010, 123) comments further that "Cities of course are more than spatial amalgams, and O'Neil's invocation of time and temperature evokes the stark conditions that accompany the desolate-to-threatening Gotham scene." It is this tie of spatiality to temporality that seems particularly significant in light of the Gothic and the type of non-linear spectral criticism practised by Jacques DERRIDA, and Jodey CASTRICANO's semiotic model of cryptomimesis. CASTRICANO defines her notion of cryptomimesis "as textual production that is predicated upon haunting, mourning, and the return of the so-called living dead" (2001, 32). The sense of haunting (as both a legacy and a promise) in the work allows it to resist lineation and CASTRICANO's theory follows the work of Jacques DERRIDA in considering the subject as phantom and the possibilities of approaching language and writing as non-linear. In *Fors* (1977) DERRIDA discusses the psychoanalytic elements of semantics, taking the symbol of the crypt beyond easy metaphors by exploring its simultaneous internal/external nature (for example as above, where it is hidden inside the word 'en*crypt*ed'), notions of absence (as seen in psychoanalytic themes whose nature is to escape from discourse), and reversal, where that which is buried alive (such as an emotion) is also in some ways satisfied by this process. As such, the object is defined as a thing to be deciphered according to a cryptographic structure, and narrative is viewed as an encrypting process.

There appears to be an implicit link between the Gothic and this kind of spectral criticism that is also seen in DERRIDA's *Specters of Marx*, which identifies a "spectral movement" (1994, 148) within *Capital* and specifically in Karl MARX's definition of the commodity (as DERRIDA equates awareness of its invisible and ideological attributes with the ability to see ghosts). Further links can also be observed when considering the ways in which contemporary Marxism has been in some ways buried and revived, and the spectral motif is even prefigured by pure Marxism: MARX and ENGELS's *The Communist Manifesto* (first published 1848) famously opens by stating it is written in response to the treatment of communism as a "spectre [that] is haunting Europe" (MARX and ENGELS 1967, 78). It seems that many ghosts haunt the texts of MARX and, through him, of DERRIDA. These approaches appear to be consistent with contemporary cultural criticism and psychoanalysis in identifying the gothic "other within" – PUNTER and BRONFEN comment that the unconscious invoked by the Gothic is not the kernel of the self but the other implanted within us (2001, 21).

DYING LANDSCAPES AND THE UNDEAD

This trope seems particularly suited to zombie narratives, and other interesting examples of the gothic and embodied landscape can be found in *The Walking Dead*, whose zombie-themed plot also seems relevant to the cryptographic model summarised above. A variety of settings feature in the story, the treatment of three of which will be discussed here: the freeway, the city (in this instance Atlanta), and 'homes' such as Woodbury, Hershel's farm and the prison.

Darren REED and Ruth PENFOLD-MOUNCE discuss the search for a haven in the AMC television series of *The Walking Dead* in sociological terms, and argue that a focus on emotional affect and mobility are the defining characteristics of the show. Kyle William BISHOP also notes the dominance of pathos in the television show, stating it "addresses the essential concerns of dramatic pathos: the struggles, losses and emotional traumas experienced by the human protagonists" (2011, 10), rather than the political or global implications of the zombie apocalypse. Many of the locations used in *The Walking Dead* are domestic (Wilshire Estates, Hershel's farm, Woodbury) and this link with the personal and familial emphasises the pathos of the story.

The issue of movement also informs the depiction of the series' setting. Stasis means death (the big cities are 'moorings' that are full of zombies) and motility (the potential for mobility) is essential. To stop moving is to die; to stay 'on the road' is to survive. This is proven time and again in both the comic book and television series, for example the ending of the second AMC television series at Hershel's farm, or the events at the prison/Woodbury in the comic. A 'stinky' RV (a literal mobile home) becomes the safest haven for the protagonists at the start of the series, and although initially only lived in by Hershel, Andrea and Amy, the entire group is soon sleeping in it, like one big family (as Rick jokes to his pregnant wife "You jogging through the *RV* to come out here and *vomit* is our new *wake-up* call," [#13]). The setting of the freeway, then, also becomes significant as these roads are quite literally the arteries of the landscape-as-body (as already noted in the context of other comics). However, in *The Walking Dead* these pathways are frequently blocked by broken-down or abandoned vehicles, and so this embodied landscape can instead be read as a dying body, whose blocked arteries and congested cities represent the country's inevitable demise. Here, gothic themes of decay and death inform the relationship between body and space, and the zombies' decaying flesh finds a parallel in the disrepair of the characters' world (see Figure 7.1).

Cities in *The Walking Dead* are the locus of a lot of zombie activity (and are thus largely avoided by the protagonists after they leave Atlanta). However, the television show contains a startling image at the end of its first episode (see Figure 7.2), as a vertical camera pullback makes the zombies frantically swarming over the tank in which Rick is trapped seem silent and remote; their movement more reminiscent of bacteria or other cells viewed under a microscope than attacking monsters. They seem part of a single organism, perhaps referencing the zombie's mindless 'mob' mentality. In a similar scene in the comic, zombies and dead bodies are heaped upon each other (Figure 7.3) and covered with flies, creating a similar impression as it becomes hard to distinguish where one body ends and another begins.

The Governor's town of Woodbury is another example of setting as bodily metaphor. When Michonne, Rick and Glenn first come across it, they comment "This *can't* be it – this place looks *dead*" (#27) and within the four fenced-off blocks that make up the town, life proceeds in a twisted facsimile of normality. Here, death and entertainment are inextricably linked through the regular fights that are held ("In case you ain't noticed, the cable's *out*. Ain't a whole lot in the way of entertainment to be had" [#27]); and, as if to reinforce this association, the Governor keeps a collection of severed heads in a stack of hollowed-out television sets/aquariums in his house, commenting "Fifty-seven channels and nothing on..." (#29). The breakdown of society is replicated in the breakdown of the characters' bodies: at Woodbury Rick loses his hand, and the Governor's abuse of Michonne results in her literally taking him apart in a torture scene that lasts an entire issue (#33).

The way in which these settings are handled is reminiscent of the Gothic, as the truth is generally revealed after a misleading introduction and a series of potential homes fail to live up to expectation. These include Wilshire Estates, Hershel's farm, West Central Prison Georgia, and Woodbury. Wilshire Estates is one of the first potential homes the protagonists come across after leaving the campsite outside Atlanta, and is presented in positive terms initially, until a cryptomimetic reveal shows that the whole estate is actually infested with zombies. Our discovery (in the final panels of #9) precedes the characters'discovery (Rick will not discover the sign until #10) as a series of panels shows the snow melting from a warning sign placed outside the estate gates (Figure 7.4). This slow reveal of a hidden truth seems cryptomimetic in nature, as sequentiality is used to reveal the sign to us across three otherwise-duplicated panels.

In a similarly gothic manner, many landmarks in *The Walking Dead* have their function inverted – for example Hershel's farm is a place of death rather than new life, where a barn is full of undead zombies rather than new crops. A prison becomes a 'safe' haven; a structure that will keep attackers out, rather than incarcerating them inside. It is described with surprise by the group "This is *nice*... with all these windows... it's not *dark* at all. I was really worried that if we moved in here we'd need torches or candles or *something*." (#14). Both functions and expectations are inverted in this manner: the prison is light not dark; the freezer in the cafeteria instead serves as a toilet ("Don't open that door! You *don't* want to go in *there*" [#14]); and the stereotypes attached to the inmates are shown to be false, for example when the 'nerd' Thomas is revealed to be a psychotic killer, literally bursting out from the interior of the prison as the truth is revealed to the group (Figure 7.5). This depiction again seems cryptomimetic – as the truth hidden inside is brought into the open and revealed. Other events at the prison, such as Tyreese's slaughter of the zombies in the gymnasium, follow a similar structure. After being locked inside with them the assumption is that he has died, until the door is opened and he is found alive, surrounded by dead zombies. As he says

"I remember it happening, but it's like I was seeing it from *outside* my body," (#16). The inversion of inside/outside recalls DERRIDA's description of the crypt as "a place comprehended within another but rigorously separate from it…sealed, and thus internal to itself" (1977, xiv), and the revelation Tyreese has survived is another event whose truth is hidden until the opening of a sealed door.

Cryptomimesis also structures other plot events and even the title of *The Walking Dead*, for example when it is revealed that all the dead reanimate regardless of how they die, as Julie's assisted suicide at the prison goes horribly wrong. Rick's consoling statement "She wasn't *bitten*. She *won't* turn, Tyreese" (#14) is instantly belied. This prompts Rick to go and dig up Shane and discover that "it's not an isolated thing – coming back without being *bitten*," (#15). This leads us towards the true meaning of 'the walking dead', as Rick confirms in his key 'We are the walking dead' speech (#24):

> You people don't know what we are.
>
> We're *surrounded* by the *dead*.
>
> We're *among* them – and when we finally give up *we become them!* We're living on *borrowed time* here. Every minute of our life is a minute we steal from *them!*
>
> You see them out there. You *know* that when we die – we *become* them. You think we hide behind walls to protect us from *the walking dead!*
>
> Don't you *get it?*
>
> We *are* the walking dead!
>
> *We* are the walking dead.

'The walking dead' can therefore be defined as a cryptomimetic term as, rather than describing the walkers, it refers to the zombie inside all of us – when we die, of whatever cause, we will join these creatures. The misleading sense (whereby it seems to describe the walkers, like George Romero's 'living dead') also supports this cryptographic meaning, as the phrase becomes something to be deciphered rather than having its meaning instantly apparent. It is significant, then, that after digging up Shane and discovering the truth, Rick shoots him, then comments "I ain't gonna bury you *again* you son of a *bitch*" (#15) and proceeds to explain the awful truth to the whole of the group.

Both the discovery and the nature of KIRKMAN's redefinition of the zombie are expressed in cryptomimetic terms and accord with the developments in the tradition more generally. The zombie has gone from possessed human (*White Zombie, I Walked with a Zombie*) to a ravenous animated corpse or ghoul (George Romero's *Living Dead* films) to infected human (Danny Boyle's *28* films; Stephen King's novel *Cell*; horror films *Pontypool* and *The Signal*).[3] This development could be linked to an increase in atheism and agnosticism and the horrors of zombification seem absent from most narratives. KIRKMAN's reboot of zombies

3 See the author's article 'The horror of humanity' (2012) for further discussion of this developing tradition and an existential analysis of *The Walking Dead*.

can therefore be read as cryptomimetic in nature (as well as reveal) as he claims there is no soul/essence to be lost through possession – we are *already* the walking dead as everyone who dies will inevitably become a zombie. As PUNTER and BRONFEN claim, the 'other' is already within us.

That this takes place in a setting full of dead and decaying machinery, blocked roads and useless technologies is a further example of the Gothic, as function is frequently inverted and the embodied landscape is mutilated and decaying. The protagonists' own bodily fragmentation through injury and violence is reflected in the destruction of the landscapes they encounter. The retrospective nature of *The Walking Dead* (as we do not see the initial zombie outbreak, just its aftermath) further defines its America as a haunted setting, where event becomes object (the apocalyptic zombie invasion was unseen, only its effects are visible). This article will now conclude by discussing the role of temporality in structuring these landscapes in further detail and in the context of contemporary comics criticism.

TIME AND SPACE ON THE COMICS PAGE

Thierry GROENSTEEN refutes the idea of breaking down the panel into smaller signifying units, focusing his analysis on the page layout, the 'macro-semiotic' level, and arguing that the system of comics relies upon three concepts: spatio-topia, arthrology and braiding, all of which require the simultaneous mobilization of the entirety of the visual and discursive codes that make up the comics page. Although he acknowledges the ultimate futility of such a debate he privileges the image over the verbal, citing the concept of iconic solidarity, whereby interdependent images are simultaneously presented as separated. His spatio-topia refers to the layout of the page: privileging space in the first instance over sequentiality/temporality. Form, area and site are the components used to analyse the panel (its shape, size and position – both on the page and within the comic as a whole). Arthrology refers to the relationships between panels and can take two forms: restrained (the sequential relationship between panels) and general (the interrelationships between all panels). Within a page's general arthrology, GROENSTEEN also distinguishes between gridding (*quadrillage*) and braiding (*tressage*). Gridding is the way a page is broken up spatially, while braiding represents the relations between any panels that supplement the coherence of the work; for example panels which appear spatially and contextually independent may still be linked through an identical construction (with differing content), a repetition of a single motif, or so on.

Both *From Hell* and *The Walking Dead* use largely regular page layouts. In *From Hell* the form, area and site of panels is constant (a 3 x 3 panel grid) in all but a few pages. In *The Walking Dead* panels are generally of a standard form (rectangular) although their size and number varies on each page, creating a more jagged rhythm where certain scenes are given more emphasis. For example, Figure 7.3 is a splash page of a single scene; whereas Figure 7.1 is a full page showing Rick's entry into Atlanta. The top panel is the largest shown, indicating its dominance, and it also functions as a background to the lower panels. Although all are regular in form, their site and area diminishes as the narrative continues. GROENSTEEN analyses 'The Plot Thickens', a strip by Bill Griffith that uses a similar device as its formal structure (each horizontal strip contains one more panel than the preceding strip). GRIFFITH's layout ironically comments on its content, as more complex and significant events are given increasingly smaller spaces of representation. Although *The Walking Dead* does not contain the sort of ironic self-awareness apparent in Griffith's title, the gridding of this page does mirror Rick's journey into the city in its use of decreasing space. It works on a simple mathematical structure. The widest and most empty space is that of the single panel that makes up the top horizontal strip (and nearly half of this panel is the white space that makes up the city's skyline), which gives way to two panels on that second strip that depict the more claustrophobic buildings, and the final strip is then made up of three panels, in which we get our first glimpse of the city's zombies.

As well as this spatial use of *diminishing* area, the temporal comes into play as the restrained arthrology of the page (the way the panels relate to each other as a sequence) works in the opposite way, *increasing* the danger by showing us more and more zombies. The three sequential panels that make up the bottom strip of the page show us first one, then two, then three zombies (in detail, with a fourth figure drawn only in outline behind). This gives the effect of a cumulative increase (but with no accompanying increase in noise or movement) while Rick becomes increasingly absent from the page as it progresses. Braiding is also apparent within the general arthrology of the page where it is used to emphasise stasis and silence. Rick's solitary speech bubble, positioned in the dead centre of the first panel, finds an echo in the "Ruh?" of the zombie in panel 4, and the absence of any other devices to indicate sound accentuates the silence. Rick's absence from the two framing panels in the bottom row also helps redefine the zombies pictured as a framing device (the two in the central panel literally frame his exit from it), and the two surrounding panels of the bottom horizontal strip frame this central one. When analysed in accordance with GROENSTEEN's system, the layout used by original artist Tony Moore on this page reinforces a reading of KIRKMAN's zombies as just another part of the decaying landscape of this still and silent world. They are an aspect of its dangers, but seldom the central peril. Instead, this page seems to suggest that isolation is the most dominant aspect, and the ultimate danger.

Analysis using both restrained and general arthrology can also inform a reading of Figure 7.5, which I analyse above as a cryptographic reveal of Thomas's true nature. On this page, the braiding between the two main panels is created through their identical form and area, as well as their iterated contents. The second panel (Thomas emerging from the darkness of the prison interior) mirrors the previous one (Andrea escaping from him); even down to the similar poses of the two characters. However, these poses are subtly inverted as Andrea's body bursts out of the building's interior, while Thomas's pose arches his body back into the blackness. The site of the second panel indicates it is the most central to our story, just as it is central on the page, and as its contents also signify (as it is here the other characters first find out who the killer is, emphasised by the reaction shot of Rick which makes up the third panel, subordinate both in terms of area and site). Figure 7.4 also uses a similar repetition of form and area to juxtapose stasis (of point of view) with movement (as time passes and more snow melts from the sign).

The concept of arthrology can be related to the temporal model that Alan MOORE puts forward in *From Hell*, which suggests that: "time is a human illusion… that all times CO-EXIST in the stupendous whole of eternity" (2.14). Events exist in stasis in 'the fourth dimension' and so time is in fact co-present and we therefore experience cyclical patterns of action throughout histor. Throughout *From Hell*, events are multiplied and revisited without recourse to a single, linear chronology, as shown in the temporal ambivalence attached to the murders and other hallucinatory scenes. Just before murdering Annie Chapman, Gull looks through a window to see a man watching television in a flat with a poster of Marilyn Monroe on the wall (7.24); he says during the murder of Liz Stride: "Yes…Yes, I remember now… This is the one that I didn't finish, isn't it? […] Has this…Has this happened before?" (8.33); his attack on Kate Eddowes in Mitre Square includes a vision of a modern towerblock at the same location (8.40); and his final murder of Mary Kelly transports him to a twentieth-century open-plan office (10.20–22).

"Let us say something peculiar happens in 1788… a century later, related events take place. Then again, 50 years later. Then 25 years. Then 12. An invisible curve, rising through the centuries," (2.15). MOORE defines a murder as the sort of 'transgressive' event that "must almost be like punching a hole through the skin of reality" (JACKSON 2000) and theorises about the creation of such a pattern by documenting a curve of murderous events as Gull visits figures such as Renwick Williams (the 'London Monster', 1788), witnesses in the 'Halifax Slasher' case (1938), Myra Hindley and Ian Brady (the 'Moors Murderers', 1963), and Peter Sutcliffe (the 'Yorkshire Ripper', 1974). This temporal model relies on a comic aesthetic, where all moments (that is, panels) are co-present in the page's spatial layout, just as all moments co-exist in the fourth dimensional 'architecture' of history. Although we humans only perceive the *restrained arthrology* of one moment to the next, relationships between the whole nonetheless exist, of the type that make up GROENSTEEN's *general arthrology*. The murders MOORE mentions are an analysis of history using this type of critical model.

Charles HATFIELD's critical model (2005) puts forward the idea that comics narratives rely upon four main tensions: between code and code (that is, word and image); single image and image-in-series; narrative sequence and page surface; and reading-as-experience versus text-as-material-object. It could be argued that these tensions are also employed in the successful creation of emotive landscape, for example in Figure 7.1, where Rick's statement "Here we are..." is juxtaposed against the empty landscape, using the tension between the visual and verbal codes to emphasise the isolation. The above analysis of the increasing sense of claustrophobia and danger can also be conveyed in terms of sequence versus page surface. While space, silence and emptiness dominate the surface of the page, claustrophobia and danger are conveyed through the sequence. The tension between single image and image-in-series is also invoked in this regard, as the alleyway Rick passes in the third panel is revealed in the fourth to contain a hidden zombie; the fifth image in turn reemphasises this by showing him continuing on his way, unaware. All these tensions are brought together in the reader's construction of the text-as-experience, which creates a temporal narrative that marches ever-forwards (like the impassive zombies!) from the space of the page.

My own observations of comics narratology have led me to argue previously for a critical model composed of three elements: the depiction of time-as-space, the involvement of the reader as interpreter and contributory author, and the creation of the hyperreal through mobile visual and verbal perspectives (2007, 316). The majority of the visual perspectives offered in *The Walking Dead* are not assigned to a character within the story, and neither extradiegetic or intradiegetic narration are used. This has the effect of allowing the reader their own identity while immersing us within this bleak world alongside the characters, although there are many points (such as Figure 7.1) where we are given bird's eye views or other angles that are strongly extradiegetic. A mobility of perspective exists herein, then; and this, along with the reader's participatory role (as interpreter and filling in the events between panels), both accord with postmodern geographies by allowing observation and evaluation from multiple points of view. My third tenet of comics narratology (time as space) also accords with some of these theorists, as follows.

Edward SOJA's (1989) argument for the reassertion of space in critical social theory goes against what he identifies as a tendency to privilege historicity over spatiality. But in comics narratology, historicisation and schematisation go hand-in-hand, as temporality can only be indicated through the use of space. Alan MOORE's tracing in *From Hell* of the "architecture" of history, that "invisible curve, rising through the centuries" (2.15), is key here. The comics medium and the content of these examples do not just support SOJA's argument, they perform and enact it. This revaluation of landscape is not only gothic, but supports the postmodern geographical critique of traditional Marxism (and its legacy) for ignoring spatiality. If postmodernism (in the Lyotardian sense) is about the rejection of grand narratives, then the landscapes and layouts of comics such as these deliver a crushing blow for postmodernism.

Here in the comics medium we can see Fredric JAMESON's prediction – that the postmodern would replace history with geography – being played out. JAMESON analyses the compass as a symbol that introduces a new dimension into our understanding of space: that of the totality. In comics theory, Thierry GROENSTEEN's *System of Comics* relies precisely on this sort of understanding of the text as a totality. But whereas for JAMESON the "aesthetic of cognitive mapping" (1992, 54) thereby becomes inherently linked to "international class realities" (52) since the subject is endowed with a "new heightened sense of its place in the global system" (54), comics offer a mediated universe (or "global system"). JAMESON argues that "cognitive mapping in the broader sense comes to require the coordination of existential data (the empirical position of the subject) with unlived, abstract conceptions of the geographic totality" (52) and he therefore concludes ideologically: focusing on historical and class-based interpretations of geography rather than its lived experience. However, I would propose that the hyperreal worlds of comics can offer a mediation of "unlived, abstract conceptions" of geography through their use of stylised art and the meaningful formal relations that exist between panels and page-parts. The depiction of the subject in comics thereby becomes one situated within a defined and meaningful landscape that is itself portrayed through the surfaces and styles of postmodernism.

The mono-sensory medium of comics, then, creates haunted landscapes that are frequently characterised by a focus upon the (spatial) effects of events. In the examples I have discussed postmodern geographies are apparent, as environment is frequently conceptualised in terms of function. Today's spaces are often defined using a critical sociological perspective and discussed in terms of their (multiple) uses: for example, even when the people (or zombies!) leave a shopping mall, it is still understood as such, containing this potential within its four walls. Perhaps, then, 'function' and space have become more like 'character' – embodied, haunted and performed. This chapter has sought to demonstrate how such an approach can also be analysed and understood as a gothic movement: relying on tropes such as abjection, fragmentation, inversion and cryptomimesis to create embodied landscapes whose qualities often reflect upon the dichotomy between the visible/invisible. Using such strategies, comics' mono-sensory narratives force their spatial depictions of the past, present and future to be thematically linked with an emotive use of space.

REFERENCES

ABRAHAM, N. and M. TOROK (1986): *The Wolf Man's Magic Word: A Cryptonymy*. Minneapolis.

AMC (2011): The Walking Dead. *AMC Series – The Official Site*. http://www.amctv.com/originals/The-Walking-Dead/ (accessed February 1, 2012).

BISHOP, K. W. (2011): The Pathos of *The Walking Dead*. J. LOWDER and J. R. LANSDALE (Eds.): *Triumph of the Walking Dead*. Dallas, 1–14.

BOICHEL, B. (1991): Batman: Commodity as Myth. R. PEARSON and W. URICCHIO (Eds.): *The Many Lives of the Batman*. New York, 4–17.

BOTTING, F. (1996): *Gothic*. London.

CASTRICANO, J. (2001): *Cryptomimesis*. Montreal & Kingston.

COPPIN, L. (2003): Looking inside out. The vision as particular gaze in *From Hell* (Alan Moore & Eddie Campbell). *Image [&} Narrative*. http://www.imageandnarrative.be/inarchive/ uncanny/lisacoppin.htm (accessed May 8, 2013).

DERRIDA, J. (1977): *Fors*: The Anglish Words of Nicolas Abraham and Maria Torok. N. ABRAHAM and M. TOROK (Eds.): *The Wolf Man's Magic Word: A Cryptonymy*. Minneapolis, xi–xlviii.

DERRIDA, J. (1994): *Specters of Marx*. P. KAMUF (Trans.). New York.

GAIMAN, N. (1989): When is a Door: The Secret Origin of the Riddler. *Secret Origins Special* #1. New York.

GROENSTEEN, T. (2007 [1999]): *The System of Comics.* B. BEATY and N. NGUYEN (Trans.). Jackson, MS.

HATFIELD, C. (2005): *Alternative Comics.* Jackson, MS.

JACKSON, K. (2000): Old Moore's ripping yarns: Alan Moore's ambitious and soon-to-be-filmed comic book novel traces the roots of the 20th century to Jack the Ripper. *The Independent.* http://www.independent.co.uk/arts-entertainment/books/features/old-moores-ripping-yarns-709999.html (accessed April 12, 2012).

JAMESON, F. (1992): *Postmodernism: Or, the Cultural Logic of Late Capitalism.* London.

KIRKMAN, R., C. ADLARD and T. MOORE (2003–present): *The Walking Dead.* Berkeley, CA.

MACDONALD, H., and P. SANDERSON (2006): New York Is Comics Country. *Publishers Weekly.* http://www.publishersweekly.com/pw/print/20060130/2746-new-york-is-comics-country-.html (accessed April 5, 2012).

MARX, K. and F. ENGELS. (1967): *The Communist Manifesto.* S. MOORE (Trans.). Harmondsworth.

MOORE, A. and E. CAMPBELL (1998–99): *From Hell.* London.

MORRISON, G. and others (1996): *The Invisibles: Say You Want a Revolution* 1, 1–8, London.

MORRISON, G. and others. (2000): *The Invisibles: Kissing Mr Quimper* 2, 14–22, London.

MORRISON, G. and others. (2002): *The Invisibles: The Invisible Kingdom* 3, 12–19, London.

PUNTER, D. and E. BRONFEN (2001): Gothic: Violence, Trauma and the Ethical. F. BOTTING (Ed.): *The Gothic.* Cambridge, 7–22.

REED, D. and R. PENFOLD-MOUNCE (2011): 'The Zombification of the Sociological Imagination: *The Walking Dead* as Social Science Fiction'. Unpublished conference paper presented at *Zombosium*, Winchester University, October 28, 2011.

ROUND, J. (2007): Visual perspective and narrative voice in comics: redefining literary terminology. *International Journal of Comic Art* 9 (2), 316–329.

ROUND, J. (2008): London's Calling: Alternate Worlds and the City as Superhero in Contemporary British-American Comics. *International Journal of Comic Art*, 10 (1), 24–31.

ROUND, J. (2012): Gothic and the Graphic Novel. D. PUNTER (Ed.): *A New Companion to the Gothic.* London, 335–349.

ROUND, J. (2012): The Horror of Humanity. W. YUEN (Ed.): *The Walking Dead and Philosophy.* Chicago, 155–166.

SOJA, E. (1989): *Postmodern Geographies.* London.

STALLYBRASS, P. and A. WHITE (1986): *The Politics and Poetics of Transgression.* London.

URICCHIO, W. (2010): The Batman's Gothic City™: Story, Ideology, Performance. J. AHRENS and A. METELING (Eds.): *Comics and the City.* New York, 119–132.

Chapter 8 | *Michael Goodrum*

THE BODY (POLITIC) IN PIECES: POST 9/11 MARVEL SUPERHERO NARRATIVES AND FRAGMENTATION

> Trauma is narration without narrativity – that is, without the ordered sequence we associate with narratives. Images are repeated but without meaning; they do not have a clear beginning, middle and end.
> (E. Ann KAPLAN 2001, 24)

Superhero narratives are not generally renowned for their sophistication. Often seen as inherently conservative and simplistic, both in form and politics, superhero comic-books are derided for offering reductive, violent solutions to complex problems. Throughout their history, there is clear evidence of superhero narratives doing precisely this but that does not mean that it is the only response of which the medium is capable. Despite the fact that initial superhero responses to the terrorist attacks in New York of September 2001 seemed to resonate with these expectations, stressing a desire for unity and emphasising the masculine heroism of rescue workers, subsequent narratives were more complex and sought to interrogate both real-world responses to trauma and the expectations of the comic-book form. One of the main devices mobilised by creative teams at Marvel in order to achieve this goal was the concept of fragmentation and their engagement with this will form the basis of this chapter.

A particular irony of this fragmentation was that it arose through an unprecedented number of crossover narratives; that is, narratives that seek to unify the storylines of a publisher's 'universe' or story world. Four such crossover narratives will be analysed in turn: *Avengers: Disassembled* (2004–5), *House of M* (2005), *Civil War* (2006–7), and *Secret Invasion* (2008). Disruption of narrative is an unavoidable part of the serial superhero format as readers are only able to access roughly twenty pages of story per month. When this occurs in a universe with multiple co-existing, loosely connected narratives, fragmentation is intrinsic to the experience of superhero narratives. In attempting to unify these disparate narratives, fragmentation is simultaneously diminished and magnified; diminished in the sense that the universe becomes increasingly interwoven and magnified as individual events become so disruptive that their effects are writ large across every single narrative. Just as in reality, 9/11 therefore becomes a temporal marker, a dividing point between past and present that interrupts the smooth flow of time. Just as 9/11 rhetoric emanating from the White House and other official sources sought to overcome the disruptive impact of trauma through an attempt to unify disparate elements, most notably through the use of such slogans as 'we are all New Yorkers', the process of uniting all of a publisher's narratives brings projects of order and disorder into conflict. What effects Spider-Man, for instance, now has an impact on everyone else – characters are no longer insulated from another's trauma, and fragmentation in one area inevitably calls into question ongoing projects of unity and order raised through the sudden proliferation of crossover narratives. Crossovers themselves, however, cannot be seen as a purely unifying project – they are a disruption to conventional narrative forms and as a result their attempts to create order and unity implicitly acknowledge the effects of trauma.

In essence, then, the presence of fragmentation both within the narrative and the form that it takes acknowledges the foundational nature it must play in these superhero narratives, acknowledging the desire for, and impossibility of, a post-fragmentation world. The origin story, that bastion of the superhero narrative, is another case in point. Often based on a traumatic event (the murder of Batman's parents, or Spider-Man's Uncle Ben, the death of Superman's home planet), the trauma provides a basis for identity – a life dedicated to fighting crime as a result of a particular event – and a life that constantly works towards the impossible, the creation of a world free of trauma, a world that returns to a time before the trauma that created the superhero. For ZIZEK (1999, 23), this would be characterised as a symptom, a "paradoxical element" that, although necessary for the conception of something as a "rational totality," is also the point at which that structure begins to break down. Superhero identity founders on the very point that created it; crossover narratives, by disrupting conventional procedures and creating an even more significant event, acknowledge the fragmentation their narrative strategies seek to disavow.

Marvel's responses to 9/11 acknowledge attempts to reconstruct new totalities but also the elements of these that operate as symptoms in the Zizekian sense. In each case, collective trauma comes into play, a formulation that takes trauma beyond the domain with which it is usually associated, the individual. It is necessary, then, to have a collective approach to trauma, something offered by Kaja SILVERMAN. SILVERMAN (1992, 55) proposes the category of historical trauma, or "any historical event… which brings a large group of male subjects into such an intimate relation with lack that they are at least for the moment unable to sustain an imaginary relation with the phallus," a process that disrupts the standard processes of patriarchal ideological investments, calling the entire system fed by these assumptions into question. SILVERMAN (1992, 30) refers to the system supported by these processes of ideological investment as the dominant fiction, or the "images and stories through which a society figures consensus" and maintains a particular set of power relations. Popular culture is explicitly identified as existing in a dialogue with the dominant fiction, drawing on the bank of representations contained within the system but also contributing new or updated versions of those representations as and when appropriate or necessary.

It is also important, now that we have a definition of collective trauma, to consider which collectives are affected. That internal processes of ideological investment in the US were disrupted by 9/11 is clear enough but it is also necessary to question the effect of 9/11 on global processes of ideological investment in the US: 9/11 was not just a challenge to the leadership in the US, it also called into question the leadership of the US. Post-9/11 superhero narratives would therefore operate in a climate where a great deal of the unquestioned principles that had characterised their adventures for decades had lost the illusion of certainty and permanence they once possessed. This would have an impact on both characters and the form of the comic books they inhabited.

AVENGERS: DISASSEMBLED

The Avengers are the leading superhero team in the Marvel universe. The crossover arc *Avengers: Disassembled* (2004–5) covered *Avengers* #500 to #503, with an additional 35 titles across the Marvel universe tying into the events in those four comic-books; the series would see the Avengers subjected to such enormous strain that they were forced to disband.[1] Ideas of fragmentation were directly acknowledged in two prominent ways – first, 'disassembled' engages with the Avengers rallying cry of 'Avengers, assemble!' and second, the Avengers symbol that appears on the front of every issue of the comic-book was smashed, as if by some violent blow. The fragmentation of the team is therefore represented in both the inversion of its rallying cry and the destruction of its symbol and can be seen to stand in for representations of the US both in itself and its position in the global order. Fragmentation also impinged on the crossover element of the narrative. Brian Michael BENDIS, the author of *Avengers: Disassembled*, stated that creative teams throughout the Marvel universe were given the option to take part in the crossover but were not obliged to do so. As a result, some teams declined to reorder their planned narrative trajectories to fit with BENDIS's ideas and in doing so increased the level of fragmentation in the Marvel universe. Read as a whole, the stories from this point in the Marvel universe are rather confusing as characters are acting in different ways and following different character arcs across a range of comic books. The 'tapestry' to which BENDIS (2004a, 36) compared *Avengers: Disassembled* at the time increasingly came to resemble a number of disparate threads rather than a coherent whole.

The preamble for *Avengers: Disassembled* explicitly engages with the concept of a traumatic return by stating that "it begins with the return of a team member thought dead" (BENDIS, 2004a, back cover). This return is realised when Jack of Hearts, thought to have killed himself while still an active Avenger, reappears at Avengers Mansion in New York.

When confronted by his fellow Avengers, the monstrous, undead Jack of Hearts is only able to utter "I'm sorry" (BENDIS 2004a, 5) before triggering an explosion that strips the flesh from Ant-Man's bones and destroys Avengers Mansion. As well as a return of the repressed, Jack of Hearts' re-emergence draws

1 For further reading on this crossover arc, see Philip Smith & Michael Goodrum, 'We have experienced a tragedy which words cannot properly describe': Representations of Trauma in Post-9/11 Superhero Comics', Literature Compass 8.8 (2011).

parallel with Slavoj ZIZEK's comments on the figure of the *Muselmann* of Auschwitz. Characterised as "the living dead, the desubjectivised subject," ZIZEK (2002, 139) positions the *Muselmann* as "the witness of what one cannot bear witness to," a statement that could be applied to Jack of Hearts in different ways. First, Jack of Hearts is bearing witness to his own death, something that necessarily occurs after the event. The undead, zombie-like incarnation of Jack that visits Avengers Mansion means that he is able to bear witness to his earlier death but the other Avengers are also forced to acknowledge their role in Jack's demise. Jack's apology can be seen as applying to his act of suicide but also the act of violence he is about to commit – a second death of which Jack has foreknowledge. Second, Jack's attack on Avengers Mansion could be seen as acknowledgement of the Avengers' complicity in the violence directed at them and by extension the US as a whole. This is not to suggest that, as some short-sighted commentators on 9/11 intimated, that the victims 'got what was coming to them', but that the Avengers, as with the US, had played a part in constructing a world order where acts of extreme violence are a matter of course and that this would, at some stage, *come back to haunt them*; in the starkly binary rhetorical world constructed by President Bush in the wake of 9/11 of victims vs. terrorists, such considerations were beyond the realms of bearing witness. Longtime Avenger Hawkeye does, however, use the phrase "we had it coming" but the rest of his statement counteracts this: Hawkeye states that the attacks came about because of the Avengers' inability to see beyond the short term, the fact that they take action in order to solve "whatever is right in front of us" (BENDIS 2004b) without thinking about the precarious nature of the position they are constructing. Effectively, then, Hawkeye is bearing witness to the myopia of previous policies pursued by the Avengers, engaging in Chalmers JOHNSON's (2001, 8) notion of 'blowback', or the "unintended consequences" of American actions overseas. In not considering the potential for their actions to foment future problems, the Avengers act in a similar way to that set out by JOHNSON.

In the disorder that follows the attack by Jack of Hearts, a more straightforward parallel to 9/11 is apparent when the Avengers' jet crashes into the remains of the mansion. Further explosions rend the New York skyline and the Vision, the android pilot of the crashed jet, informs his fellow Avengers that they are "no longer in control of anything that we, as a group, held dear" (BENDIS 2004a, 500 before he, too, dies having reiterated that the attack comes from within the team itself – that the Avengers, in Vision and Jack of Hearts, include what amounts to two suicide bombers. Conventional frameworks are fragmented as heroes become both villains and victims – neither Jack of Hearts nor the Vision want to do these things, they are compelled to do them by a higher power – and this can be seen as a commentary on the US after 9/11: guided by a higher power (the need to combat the communist threat), the US did things it may not have wanted to do during the Cold War; 9/11 was a response partially generated through the forces unleashed by these earlier actions, one for which complicit elements, formerly seen as heroic, are sorry.

After the attacks by Jack of Hearts and the Vision, the Avengers are attacked by sinister robots and the heroic Wasp is seriously injured; before the Avengers can recover, an alien invasion fleet appears over New York and, in the ensuing battle, a badly wounded Hawkeye sacrifices himself to destroy the command ship – the third instance of an Avenger as a suicide bomber, albeit one framed as heroic sacrifice. A sequence of such apocalyptic events arouses suspicions that all is not right, suspicions that are confirmed by Dr Strange, Earth's sorcerer supreme. It would appear that the forces of magic that flow through the world (a little like the 'force' in *Star Wars*) have been abused and there is only one person capable of exercising such power – the Scarlet Witch, a current Avenger. Her actions have been motivated by trauma – by another return of the repressed. For many years, Scarlet Witch had been married to the Vision and, thanks to her magical powers that allow her to influence probability and ultimately reality, managed to have two children despite the fact that she was married to a robot – two children, it transpired, that were not real. Rather than individual beings, Scarlet Witch's children were parts of the soul of the demon Mephisto and, when these parts were identified and re-assimilated, the children disappeared. Scarlet Witch's memories of the children are then suppressed so that she can avoid the traumatic impact of this loss; something she succeeds in doing until she recovers her memories and vows to take revenge against those who deprived her of her children through orchestrating the bombing, crash and alien invasion. When this becomes apparent, she is captured and subdued and the Avengers are left to deal with the destruction she has left in her wake. Ultimately, the team disbands.

Avengers: Disassembled is therefore characterised by its association with trauma and fragmentation. Flagged by the name of the arc and the shattered Avengers logo, fragmentation comes to define almost everything about the narrative. Families are divided by death; the Avengers come to an end; bodies and buildings are subjected to a strain greater than they can bear. The lack of superhero control is summed up in *Avengers* #500 when, after Jack of Hearts's attack on the mansion, Hawkeye is told to take point – "point of what?" is his response (BENDIS 2004a, 7). Unable to offer a centralising force, superheroes prove to be just as susceptible to the impact of trauma as everyone else. Destruction is a recurrent theme in superhero narratives but it does not usually lead to such disorder in the superhero ranks. A commonly repeated phrase throughout *Avengers: Disassembled* is "what is this? Why is this happening?" and these seem to be questions that do not normally occur to superheroes. Despite the fact that the nature of the destruction resonates with 9/11, it is the response that is of greater interest. The inability to process the events occurring around the Avengers, demonstrated through the frequent reappearance of a particular phrase, recalls the Freudian ideas of preparedness and repetition.

When it comes to repetition there are a number of things to consider. First, there is the linguistic repetition already identified. Second, there are repeated flashbacks, themselves indicative of trauma, that also serve to both fragment the narrative and also drive it in circles as the characters look for the meaning of these flashbacks. To further complicate the structure of these intrusions from the past, the reader takes a subtly different meaning from each retelling as context changes and further knowledge is accrued. Flashbacks in *Avengers: Disassembled* are usually traumatic, with Captain America frequently being returned to the moment when Bucky, his youthful sidekick, died during the Second World War. Suggestive of a "compulsion to repeat" that, for FREUD (2001, 22), overrides the conventional way in which the mind works, these instances of traumatic return are too much for Captain America to bear; he is in a passive position, repeatedly subjected to unsettling flashbacks that do little to add to his mastery of events. Repetition of this type goes against what may be expected from superhero narratives, particularly given WOLF-MEYER's (2003, 510) statement that they are "predicated on preserving the status quo" and are not open to interpretation. In fact, WOLF-MEYER states that "the majority of comic-book readers are limited to a specific reading" and that "there is very little room for interpretation given to them by the authors of the text," (2003, 510) an approach that runs into increasing difficulties as the post-9/11 Marvel narratives progress. In *Avengers: Disassembled*, superheroes and their actions are unable to fulfil the conservative project of retaining the status quo (a veiled commentary, perhaps, on the ability of the US to restore or reinstate a world order where they continue to enjoy the same privileges as they had before 9/11) – they are also open to interpretation; in fact, they seem to welcome it through the repetition of such language as 'what is this?'

Avengers: Disassembled ends with the surviving members of the team back in the wreckage of the Mansion. It is here that they all announce their departure from the team. Iron Man states that the derelict mansion will be left to stand as a memorial. Unlike conventional superhero narratives that end with the heroes triumphant and at least temporary narrative resolution, the final issue of *The Avengers* is a fragmentary affair. In terms of narrative, it is a series of reminiscences of team members' favourite events; in terms of form, the art style is fragmented as it is drawn by a number of different artists, all of whom have markedly different styles. The piece does not, therefore, hang together as a whole as convincingly as usual.

There is, however, one particular demonstration of unity worthy of comment. As the erstwhile Avengers leave the mansion they are greeted by a large crowd united in grief, recalling scenes from New York in the days and weeks after 9/11; in terms of the dominant fiction, this is a complex display of solidarity. Superhero narratives have a predominant tendency to end with the hero triumphant – the instance of trauma that precipitated their initial involvement has been resolved and the pre-traumatic status quo re-established albeit, given the serial nature of the superhero format, tentatively. A mass demonstration of grief, while running contrary to the theme of fragmentation, does not necessarily reinforce the dominant fiction in any positive sense. As a result, although the grieving gathering demonstrates collective unity, it does little to shore up the mechanisms that work towards sustaining that unity. Ideological investment in the dominant fiction, or at least the pre-traumatic construction of the dominant fiction, is indicated but, unlike a triumphant ending, there are no attempts to resolicit ideological belief and to establish the dominant fiction as an ongoing, permanent, construction. Instead of closure, the narrative bleeds into another crossover: *House of M.*

HOUSE OF M

House of M (2005) follows on directly from *Avengers: Disassembled* and engages with the aftermath of the earlier arc, particularly the mental breakdown of Scarlet Witch. A stand-alone series of eight issues in its own right, *House of M* was more ambitious in its approach to crossover than *Disassembled*, covering 61 separate comic-books and sparking specific tie-in series such as *House of M: Avengers* #1–5; unlike *Disassembled*, *House of M* had an impact on every narrative in the Marvel universe. It begins with Charles Xavier and Emma Frost, Marvel's leading telepaths, debating how best to control Scarlet Witch. Neither of them can be sure that any mental mechanisms put in place would be able to contain the traumatic memories that triggered the Scarlet Witch's previous actions and, as a result, Emma Frost suggests that the only viable option is to kill Scarlet Witch. Quicksilver, Scarlet Witch's brother, overhears this and instantly goes to his sister; together they devise a plan to prevent her from being killed. The plan itself is perfectly simple yet indicative of Scarlet Witch's power – using her control over reality, she will fabricate an entirely new reality, one in which everyone will have their heart's desire and no recollection of the reality that had gone before.

House of M therefore testifies to the constructed nature of reality. In so doing, it raises questions about ideological investment and the ability of the dominant fiction to establish itself on a permanent basis. Scarlet Witch's dissatisfaction with the existing reality is easy enough to understand – not only have her children been exposed as fictional and destroyed, she is responsible for the death of her husband (the Vision) and her friend (Hawkeye) and she is also facing the threat of death as a result of these actions. It is therefore clear why she would find her new reality preferable; what is, perhaps, more remarkable is that as the narrative progresses other superheroes come to agree with her.

The newly constructed reality is, to some extent, an inversion of the one that pre-existed it. Mutants, beings possessing a specific gene that makes them different from humans, grants them certain abilities and has often led to their persecution, are the rulers, with humans the oppressed minority; this represents the desire of Magneto, Scarlet Witch's father. Other characters seem to slot nicely into this construction, apart from the human superheroes, the bulk of whom form the core of a resistance movement. There is only one exception to the norm of mutant happiness: Wolverine. Whether due to the effects of trauma or some deliberate mental block imposed by external sources, Wolverine has never been able to remember his early life; his desire to do so, in fact, has contributed to the dynamics of his character arc for several decades. Scarlet Witch fails to take this into account when constructing her reality and therefore she makes Wolverine's repressed memories available to him. In so doing, the constructed nature of the new reality is exposed and subjected to assault, both physical and ideological. Recovery of memory, the reason for the attempt to create a new 'rational totality', therefore comes to operate as a symptom, necessary for the new social order but also the point at which it begins to unravel.

Fragmentation and trauma play a significant role in *House of M*. The fact that the narrative thrust of the story is derived from a previous story line engages with the idea of trauma as either "the encounter with death or the ongoing experience of having survived it," an approach that identifies trauma as both a response to, and an ongoing process of, living in relation to an event (CARUTH 1996, 7). The construction of *House of M* therefore engages with CARUTH's definition of trauma as "the narrative of a belated experience, [that] far from telling of an escape from reality... rather attests to its endless impact on a life" (1996, 7). *House of M* demonstrates this endless impact rather effectively by highlighting that the effects of trauma are inescapable – even a retreat into a carefully constructed fantasy cannot fully insulate the traumatised individual from the effects of their experience.

Equipped with the knowledge that he is living in a fantasy, Wolverine sets out to challenge Scarlet Witch and the reality she has constructed. In order to do this successfully Wolverine has to recruit allies. The only problem is that, due to the false pasts provided by the new construction of reality, none of them are aware of their past adventures with Wolverine. Unable to convince them alone, Wolverine is assisted by a young girl who, with the advent of the new world, discovers she has a mutant ability – to tell when reality has been altered and to demonstrate the fact to people in close proximity. What follows are a series of flashbacks, explicitly rendered as traumatic, that subject both the individual experiencing them and the fabric of Scarlet Witch's reality to increasing strain. By re-experiencing traumatic events, the lasting impact of trauma is also established; trauma is responsible for both the construction and the fragmentation of Scarlet Witch's false reality.

Strain is visualised in *House of M* through both flashbacks and the impact of trauma on the body. Both lead to fragmentation, whether of narrative or the body itself. Fragmentation of the body recalls the Lacanian concept of the body in pieces where, according to LAPLANCHE and PONTALIS (1973, 251), "anxiety about fragmentation can at times be seen to arise as a consequence of loss of narcissistic identification." First, fragmentation itself becomes a narrative trope with the dissolution of individuals standing in for the wider assault on society. Second, fragmentation occurs in different ways and under different circumstances. There are other aspects to consider but these two areas will form the basis of discussion.

Wolverine's attempts to gather his old friends connects with a predominant theme of trauma studies, the recovery of memory, with the instances of remembering explicitly represented as traumatic. Past scenes of loss and violence interrupt the narrative, subjecting the individual to mental trauma that is often represented physically – those affected appear visibly shaken and the ways in which their bodies are represented differ considerably from superhero convention. Instead of the confident controllers of their environment, the traumatised superheroes are slumped, overpowered by both their surroundings and their recovered memories. As if to draw attention to the fact that something is amiss, the gutters between the panels, usually left clear in an attempt at invisibility, are filled in with black; a creative decision of this nature serves to highlight both the sombre mood of the piece and the void behind the construction of reality of which, as a result of Scarlet Witch's magical alterations, many of the superheroes are painfully aware.

Peter Parker's flashback is the longest and most detailed of any superhero in *House of M*, stretching across four pages. The page above best demonstrates the physical effects of the flashbacks and also takes into account Parker's recall of the death of Gwen Stacy, his first girlfriend, and Uncle Ben, his fatherly guide. Despite being more detailed than others, Parker's response is relatively typical and this is understandable as the spaces that surround the superheroes are a product of, as well as a response to, traumatic events. In their perception, the entire world in which the superheroes find themselves is a hostile, oppressive environment that attests to a fragmentation of both their identities and the fabric of reality. Just as there is a false reality, there is a false back story for each of the characters and, despite the fact that these are proved to be fictitious constructions, emotional attachments are all too real. The act of recovery is therefore also an act of loss.

The impact of these fabricated emotional attachments is most clearly demonstrated in the case of Spider-Man. In continuity – that is, in the 'real' comic book reality – Gwen Stacy died, as did her father and Parker's Uncle Ben. In Scarlet Witch's world, Parker is married to Gwen and both male relatives are still alive – Gwen and Peter also have a son, as can be seen in Figure 8.2. Despite the fact that Parker is convinced of the fact that the reality in which he finds himself does not really exist, the emotional attachment he feels to his wife and family cannot be so easily dissipated. Parker is therefore forced to relive his trauma and also add to that through the loss of a son. Parker's trauma therefore doubles that of Scarlet Witch, suggesting that despite her herculean efforts to avoid her trauma, Scarlet Witch is condemned to repeat her unpleasurable experiences; not only that, but the trauma of the Scarlet Witch has become another symptom, a fundamental part of her constructed reality representing another point where the social order is vulnerable.

The overpowering nature of trauma is also demonstrated in the case of Brian Braddock, ruler of the multiverse and guardian of reality. As Scarlet Witch's actions constitute an abuse of, and a threat to the continuing existence of, reality, Braddock enters the fabricated world with the intention of destroying it. Once Braddock comes into contact with this false world, though, he is overpowered by it and has no recollection of his mission. The idyllic life Braddock finds himself living in his ignorance is subsequently punctured by traumatic nightmares.

Trauma is clearly evident in the figure above. The images themselves are harrowing, with the central image in particular suggestive of the body in pieces; the way in which they are presented also suggests an engagement with trauma. Instead of the usual ordered panels, those in Figure 8.3 are odd shapes with jagged edges that cut into each other and fail to fill the page. Braddock's dream engages with Susan FALUDI's (2008, 12) statement that "the successful dream orders experience; the nightmare confounds order, alerts the sleeper that the wished-for narrative isn't holding" and this becomes particularly apparent when the nature of the panels is considered. Like the reality in which Braddock finds himself, the panels seem to be collapsing under the impact of trauma, not providing order but rather a platform for disorder that hints at the constructed nature of the reality depicted and the comic-book itself. Gaps between panels are generally seen to account for transitions in time and space but here the gaps, the absences, hint at a process of construction. In bearing witness to the white space behind the panels, audiences are jolted out of their usual reading strategies and reminded that what they are reading is not real. Trauma therefore disrupts form, content and conventional reading strategies.

One aspect of the narrative apparently immune to change in *House of M* is the need for superheroes to resolve their difficulties through violence. Despite reservations on the part of some, most notably Spider-Man (who does not want to lose his family) and Hawkeye (who was killed in continuity), Wolverine and his allies mount an attack on Scarlet Witch. Instead of the straightforward victory so often secured in superhero narratives, the battle itself is marked by fragmentation and disorientation and its aftermath is far from triumphant. In the initial assault, the Scarlet Witch fragments in the arms of her father, leading him to repeat a phrase from *Avengers: Disassembled,* 'What is this? What is happening?' Magneto's statement forges a link to the earlier arc, indicating an ongoing inability to process traumatic experience and simultaneously engaging with the idea of repetition as central to trauma. In fact, repetition plays a central part in the analysis of fragmentation at this point.

Throughout *House of M,* Scarlet Witch has been represented as fragmenting from the head downwards; the same is true in the fight scenes that constitute the climax of the arc. As a result, the mental problems at the base of Scarlet Witch's fragmentation are emphasised – she is unable to 'pull herself together' and therefore quite literally falls apart, taking her reality with her. When it comes to the fragmentation of Hawkeye, however, the process is noticeably different.

The fact that Hawkeye is in a position to be fragmented indicates a traumatic return. Like Jack of Hearts before him, Hawkeye has returned to bear witness to the impossible, to his own death. Scarlet Witch, already responsible for Hawkeye's death in continuity, apparently has no choice but to repeat her actions when confronted with Hawkeye again in *House of M* #7. Crucially Hawkeye, like Scarlet Witch and her reality, fragments and falls apart; unlike Scarlet Witch, the continuing presence of Hawkeye's head suggests a desire for acknowledgement and continuity of being – a coherence of identity absent from Scarlet Witch.

After Scarlet Witch is forced to confront her traumatic past, her constructed reality falls apart and the superheroes are returned to their previous reality. As with *Avengers: Disassembled,* the arc does not end on a positive note. Some of the superheroes, most notably Spider-Man, are able to remember their false lives and are traumatised by their losses. The superhero community is therefore divided into those who can and cannot remember these events, a source of fragmentation in itself, but there are also arguments when Doctor Strange, master of the mystic arts, tells Spider-Man that he cannot remove the painful memories. Superheroes end the arc fragmented, at odds with one another and lacking a clear purpose. In terms of the dominant fiction, superhero narratives are doing very little to solicit ideological belief in the continuation of the existing system. Effectively what the pyrrhic victory over Scarlet Witch suggests is that while the current system is worth fighting for, no-one is entirely sure why or what benefits it has for those within it. The nature of the system, and its merits, would be explicitly discussed in *Civil War*.

CIVIL WAR

As *Civil War* (2006–7) begins, a group of young, inexperienced superheroes try to take on an older, more organised team of villains. In the battle that follows, a large part of an American town, Stamford, is destroyed and 600 people are killed, including a large number of school children. A backlash follows and there are calls for legislation to control superhero activities; rather than showing a united front in the face of this challenge, those with superpowers divide into competing factions. The most notable groups are those led by Captain America, who oppose the Super Human Registration Act (SHRA), and Iron Man, who campaign on its behalf. Whereas the two previous arcs featured threats to the existence of the world, *Civil War* deals with a threat to the existence of the world as we know it – a political challenge that cannot simply be met with violence. Like *House of M* before it, *Civil War* was both a stand-alone seven issue arc in its own right and also fed into every narrative in the Marvel universe, covering over 100 contemporary comic books.

Traces of the previous arcs continue to manifest themselves, most notably through the loss of children. The death of Scarlet Witch's children proved the foundational trauma for *Disassembled* and *House of M* (doubled in the loss of Spider-Man's son) but in *Civil War* the children of Stamford are mobilised by the media as proof of the fact that superheroes, in their current organisational form, are incapable of protecting those who need it most. It is therefore necessary to consider what it is these children symbolise and why they have come to routinely serve as a structural prop in post-9/11 Marvel superhero narratives.

Significant capital, both cultural and financial, has been extracted from comic-book storylines declaring the death of a leading character. The first instance of the death of a superhero attaining levels of mass cultural saturation was in 1992 with *The Death of Superman* – a narrative device so successful that the trade paperback collecting together the original comic-book narratives bears the legend 'the best-selling graphic novel of all time!' on its front cover. Aware of the gains to be made by accessing this discourse, *The Avengers* #502, part of the *Disassembled* arc, had 'one of these Avengers will die!' emblazoned across its front cover. When it comes to children, however, there is little demand to read of their deaths. In choosing to emphasise the deaths of children, then, *Civil War* engages with concerns raised about the ability of security forces in the US to protect their most vulnerable citizens. This structural feature, gradually increasing in importance up to this point, could therefore be seen to suggest a shattering of faith in the dominant fiction to sustain itself and its relationships of power; in this respect, the death of children could also be seen as the loss of the future as an extension of the present. The child represents continuity from its parents – his or her death indicates a radical dislocation, a break from the past that calls into question ideological belief in existing institutions. Child fatalities could also be seen as an end of innocence, a return from ironic, playful postmodernity to the Real made explicit through shocking violence; in itself, however, this is ironic given that it occurs in a medium whose violence has long been regarded as childish, immature – as an escape from reality.

A more explicit focus of attention in *Civil War* is legislation that covers what people are able to do and how they are able to go about these things; for this, and other reasons, the arc has been seen to mediate discussions of the USA PATRIOT Act. Although apparently engineered to promote complicity through its laboriously crafted name (Uniting and Strengthening America by Providing Appropriate Tools Required to Intercept and Obstruct Terrorism), the PATRIOT Act inspired resistance from civil liberties groups who objected to, among other things, the enhanced powers of surveillance granted to the government that, according to Nancy CHANG (2002, 13), threatened "to intrude upon the privacy of everyone in the United States." *Civil War* therefore engages with challenges to a system from within while also trying to set out, through dialogue, why certain systemic features are more desirable than others. To a significant extent, this dialogue is dominated by those who seek to gain rhetorical capital from the dead children of Stamford, most explicitly demonstrated when Iron Man ends a debate with Captain America by stating that he should address his civil liberties concerns "to the parents at Stamford" (HUDLIN 2006, 19). An approach such as this finds favour with much of the public, already whipped up into a frenzy by the media, but parts of the superhero community fail to be easily convinced and comparisons are drawn to both Nazi persecution and racial discrimination.

Civil War therefore takes the relatively unusual step for superhero narratives of calling into question, or fragmenting, the ideological investment of superheroes in the dominant fiction. Action in *Civil War* is driven by ideological conviction, something that is set out through extended conversations; there was even a dialogue-heavy forerunner, *Road to Civil War*, which explicitly framed the coming arc as being about a community divided by competing ideas. However, it is not only communities that are divided – some individuals find the conflicting arguments difficult to master and as a result are unsure of how to act. In short, there is no unified response and, despite the superheroes largely falling into two camps, those 'for' and 'against' the legislation, a range of positions are made clear within this simple binary mechanism. Ultimately, *Civil War* shatters the idea of a superhero united front.

Audience debate was actively encouraged throughout *Civil War*, most noticeably through its use of the tagline, 'Whose side are you on?' Further debate was prompted by the role of Spider-Man. One of the conditions of registration is that those who sign up have to disclose their secret identities, a course of action that Spider-Man has consistently resisted for fear of the consequences it may have for his family. As one of the first registrants, however, Spider-Man has to set an example and therefore holds a press conference to put an end to his split personalities. After learning more about the nature of the Registration Act, however, Spider-Man becomes uneasy. Iron Man, the architect of the legislation, has constructed a prison in the negative zone, a kind of no man's land between dimensions, where non-registrants will be held indefinitely until they sign. In certain respects, then, this prison mediates the US facility at Guantanamo Bay – and Spider-Man's disgust with the way in which prisoners are held outside conventional legal frameworks works towards encouraging audiences to reflect on the real world situation faced by terror suspects. Considerations of this nature were advanced through the tie-in series *Civil War: Front Line* as, at the end of each comic-book, a brief feature explicitly set out the connections between events in *Civil War* and reality – although rather than Guantanamo, *Civil War: Front Line* drew a parallel with the internment of Japanese-Americans during the Second World War, a rather less risky position. Spider-Man's inability to reconcile himself to this facility, and the wider consequences of the Registration Act, lead to him changing sides; Iron Man's forces meet this decision with extreme violence and Spider-Man is nearly beaten to death.

The attack on Spider-Man, along with an assault on hero Luke Cage for refusing to join either side, works towards promoting debate among the audience. In doing so, the narrative encourages individuals to adopt their own approach, promoting the idea of a fragmented and critical readership. Contrary to WOLF-MEYER's (2003, 510) implication of a passive readership and a narrative form that resists interpretation, *Civil War* dispenses with all pretence of a cohesive audience and acknowledges, in fact encourages, fragmentation and division. Anti-hero Deadpool's straightforward interpretation of registration, "I'm an American and I got a badge and that gives me the right to kick the crap out of anyone anywhere anytime to do what's right," with the 'right' being dictated by the US government, seems to be an attack on previous constructions of superhero narratives as unthinking, uncritical champions of the status quo (NICIEZA 2006). As indicated by Spider-Man's shift in allegiance, even the broad churches of 'for' and 'against' registration prove too constrictive. As the battle between the factions continues, growing increasingly costly in terms of damage and loss of life, public sentiment begins to shift behind the pro-registration forces who present themselves as a force of order. Displays of public opinion come to a head when Captain America is prevented from landing a final blow on Iron Man by a group of people indicative of responsible citizenship, most notably a fireman and a police officer – positions of increased ideological investment in the wake of the 9/11 terrorist attacks.

What the attack on Captain America also demonstrated is the breakdown of the relationship between the superhero and the community he protects. In the case of Captain America, this relationship is usually taken for granted; Captain America is an embodiment of the nation that exists in an unproblematic relationship of ideological investment with their champion. In the words of Jason DITTMER (2005, 27), Captain America is "a hero both of, and for, the nation," a position that becomes increasingly complex as politics become increasingly divided. In the short term, at least, *Civil War* cannot be seen as an attempt to, as DITTMER remarks, "create order out of the complexity of global events by constructing narratives through which the region's place in the world is understandable and legitimate" as the focus of the arc is on calling events and approaches into question (2005, 627). Instead, Captain America represents a specific faction of the US – and the fact that it is Captain America, the noblest of superheroes, holding this position could be seen as an injunction by the creators that this is the kind of position to which the US should aspire. In this sense, Captain America is still creating order out of complexity, but his project remains unfinished as the assassination of Captain America forms the climax of *Civil War*. Left to their own devices, the superheroes of the Marvel universe would need another crossover arc to even begin the process of reconstruction.

SECRET INVASION

In the wake of *Civil War*, a new order was created. Under the guidance of Iron Man a process called the Initiative begins, a process that leads to the creation of a centrally trained and administered superhero team in each state of the US. Many of those who had previously opposed the new system continue to do so but, ultimately, Iron Man seems to have created a powerful force for order. Appearances, however, can be deceptive and it soon becomes apparent that the superhero civil war and the Initiative form the backdrop to the first acts in an invasion by an alien race known as Skrulls. These shape-shifting aliens have been a recurring thorn in the side of Marvel's heroes for many years but have never previously attempted such an ambitious action. However, there is a clear paradigm shift when it comes to the characterisation of the Skrulls in *Secret Invasion*. Formerly aliens apparently bent on inter-galactic conquest for its own sake, the Skrulls are now identified with religious extremism through the use of words such as 'jihad' and the widespread use of religious, predominantly Islamic, terms, a decision that created a direct connection to the 9/11 attacks. The ability of the Skrulls to pass as anyone targets the bonds of trust that tie the superheroes together into effective fighting units (while acknowledging that a terrorist could be anybody), creating doubt in the minds of the heroes as to whether anyone is who they claim to be, and as a result the Skrulls come closer to success than ever before.

In the early stages of the invasion, with superheroes unable to tell if anyone is truly who they claim to be, the Skrulls make significant advances. Each of the state-based teams in the Initiative features a Skrull and as such superhero responses to the invasion are massively disrupted. As individual heroes able to identify Skrulls begin to assemble teams around them, the superheroes begin to enjoy some success; Iron Man does not share these, however, as the Stark Technology that runs almost everything in the US, including Iron Man's suit, is hopelessly compromised by the invading Skrulls. The Skrull queen also succeeds in causing Iron Man to doubt his humanity, forcing him to confront the fact that he may be a Skrull. The crumbling of Iron Man's identity, his loss of belief in himself, mediates the wider identity crisis of the order he has fashioned as those within it are unsure if anyone is who they claim to be and the structures supposed to guarantee safety, the superhero teams in each state, have been used as a vehicle for attack. With Iron Man incapacitated, other people emerge to lead the forces of Earth, most notably Norman Osborn, the man formerly known as the villainous Green Goblin, who had been released from prison and trained under the terms of the SHRA. As the forces of Earth regroup, the campaign against the Skrulls gradually gathers momentum and ultimately the Skrulls' "jihadist" attempts to claim Earth are defeated, a moment that comes to be represented by Norman Osborn shooting the Skrull queen live on a television newsfeed being broadcast to millions, an act that subsequently leads to him being placed in charge of US security forces ('jihad' is a term used in several comic-books to describe the Skrull invasion. See, for example, *Nova* #16 [2008], *New Avengers* #44 [2008] and *Guardians of the Galaxy* #6 [2008]).

Secret Invasion (2008) still offers a relatively straightforward narrative – aliens invade, they are repelled. As with the crossovers that went before it, a great deal of complexity was brought to this simple story through the fact that it tied into every other Marvel narrative covering, like *Civil War* before it, roughly 100 single issues as a result. The arc engages directly with fragmentation through one of these tie-ins, *Secret Invasion: Front Line*. Throughout *Civil War*, the *Front Line* series had been a vehicle for critical engagement with the themes raised throughout the arc; during *Secret Invasion* it would become a means of representing the effects of historical trauma. The narrative is dislocated and it is at least initially difficult to identify a central protagonist as *Front Line* follows groups of only loosely related people. In *Civil War*, the journalists Ben Urich and Sally Floyd gave *Front Line* purpose and drive with their investigation into the foundations of the Initiative, offering a narrative of crusading journalists in a similar vein to *All The President's Men* (1976). There is no such drive running through *Secret Invasion: Front Line* as the returning Urich appears lost, unable to exercise control over the events in which he finds himself. Instead of an investigation, *Secret Invasion: Front Line* is structured around playback from Ben Urich's Dictaphone.

The narratives that emerge from Urich's Dictaphone are disjointed and fragmented. As KAPLAN (2001, 204) remarks, trauma narratives are related through "means other than linearity or story [such as] fragments, hallucinations [and] flashbacks," all of which are apparent in *Secret Invasion: Front Line*. It is never clear whether Urich has captured coherent narratives on his Dictaphone as his use of the rewind function offers up only disjointed fragments; even these often have other interviewees recorded over them, adding to the levels of fragmentation. The multiple recordings also recall Spider-Man's description of the aftermath of *House of M* where he describes the state of his mind as "memories on top of memories," a phrase suggestive of a breakdown of conventional mental structures (BENDIS 2005, 18).

Visual flashbacks to *House of M*, most notably the fragmentation of Hawkeye and Scarlet Witch, position the "endless impact [of trauma] on a life" identified by CARUTH (1996, 7) in a slightly wider context. In the post-9/11 Marvel universe, trauma has an ongoing impact on many lives, not just one – a process facilitated through the repeated use of crossover arcs to draw all the characters together. An attempt to resist the impact of trauma can be seen in the decision to continually bring so many superheroes together (another explanation could be that crossover arcs are good for business). The fragmentation of Hawkeye and Scarlet Witch from *House of M* was restaged some three years after it initially appeared in order to demonstrate how long the Skrull invaders had been present on Earth prior to the military arm of their attack. Many images could have been chosen for this that would have been equally, if not more, effective in showing the passage of time. What the use of these images of fragmentation showed was the effect they had on both creative teams and audiences – even the Skrulls caught up in the events of *House of M* describe them as "traumatic… [and] too much to even say," a statement that indicates the magnitude of the trauma in question; even those planning to cause a traumatic event were unable to offer sufficient psychic resistance to the assault of fragmentation and the horror of the body in pieces (BENDIS 2008, 22).

Themes of trauma and fragmentation had previously been played out through the loss of children; *Secret Invasion: Front Line* continued this theme but in a different sense. Whereas previous arcs had dealt with the death of children, *Front Line* featured a section on the loss of a parent. Melanie, trapped with her father, Vincent, in his high-rise office when the invasion begins, is a teenager and therefore a different prospect to the very young children of previous arcs. Traumatised by Vincent's death, Melanie is unable to attend her father's funeral – suggesting that, while there is a dislocation from the past, it is one that Melanie is unable to fully acknowledge – her life still exists in relation to death. Her status as a teenager, however, places Melanie in limbo between continued parental reliance and the assertion of her independence; despite her inability to come to terms with the violent death of her father, then, this can still be seen as a position orientated towards the future, albeit currently overwhelmed by the past. *Secret Invasion* therefore closes the circle of child death narratives by offering its readers the possibility of a future, albeit one forever in debt to the past – Urich expresses hopes that Melanie will come to terms with her loss in time while images of the reconstruction of Manhattan are presented to the audience, suggesting that this process is already underway. This ties in with another child-based narrative device from *Secret Invasion* – that of the Skrulls' abduction of Luke Cage and Jessica Jones's daughter, Danielle, a move that gestures towards the destruction of the future but is defeated through her safe return to her parents.

The fact that both children are girls, however, hints at a reformulation of existing relations. These are not sons who will avenge the wrongs of the past but daughters who, if we follow the lead given to us by Urich in *Front Line*, will be involved in processes of reconstruction – in building a better future. To some extent, then, this approach hints at flaws in the dominant fiction that can be improved, flaws immediately identified by the fact that those chosen to symbolise this process are women and therefore marginalised by existing patterns of masculine dominance. Promises of revision, however, are submerged within the need to overthrow the next threat, suggesting their immediate deferral – a device that could be extended indefinitely, if needs be, until a point is reached where revisions become less desirable and the dominant fiction is able to re-establish itself more firmly. Revisionary potential therefore exists only as precisely that – potential – and its success is dependent on levels of fragmentation in two ways: first, fragmentation initially makes the moment of revisionary potential possible and second, the context of the moment shapes attitudes towards revision. Until this point, change has largely been presented as negative and the climax of this is undoubtedly the ascent of Norman Osborn to a prominent government position. Even while the comic books are opening up these spaces they are simultaneously working towards closing them off, constructing the terms of debate in favour of existing relations of power – but the glimpse of an alternative, the acknowledgement that the current framework is not the only possibility, can be seen to undermine long term projects of stability.

CONCLUSION

Fragmentation operates in different ways in post-9/11 Marvel comic books and the functions it serves are different both between and within crossover arcs. Initially, fragmentation demonstrated the potency of trauma, showing how even superheroes used to dealing with trauma could be overpowered by events. By *Civil War*, Captain America appeals for more time to work through the aftermath of the trauma in Stamford but events seem to have taken on an overwhelming momentum – Captain America is killed and knee-jerk reactions are privileged. Uncritical reactions to trauma tend not to work towards coming to terms with the events in question. In fact, decisions arrived at in such a short space of time, as Arthur G. NEAL (2005, 17) remarks, tend to result in "strong moralistic judgements of right or wrong," with little consideration of what 'right' and 'wrong', 'good' and 'evil', actually entail. What this effectively means is that, according to NEAL, events understood in such a superficial manner are placed "within such pre-existing frameworks as religious fundamentalism or a political ideology" (2005, 17); rather than being seen as individual events, they are assimilated into existing systems for making sense of the world. *Civil War* demonstrates a significant shift from the confusion of the first two arcs, characterised as they were by an inability to process traumatic events in any effective way; not towards understanding trauma, but towards mobilising it for concrete political goals. Iron Man's new political order is only able to succeed due to the way in which it engages with, or rather fails to engage with, the trauma of the deaths in Stamford. Captain America's failure and subsequent death can therefore be seen to represent the inability of civil liberties groups to overcome government rhetoric and to alert the public to its shortcomings. *Secret Invasion*, to some extent, combines approaches from all three previous arcs. The fragmented narrative in *Front Line* embodies a similar approach to trauma to that found in *Disassembled* and *House of M*, while the ascent of Norman Osborn reiterates the dangers of a superficial approach to trauma but, through placing a known villain in the position formerly occupied by Iron Man, takes criticism one step further.

There are both continuities and discontinuities in the post-9/11 Marvel approach to trauma, then. It would be misguided, though, to presuppose some form of teleological progression, whether for good or ill. As with trauma itself, these narratives are best approached in terms of moments; not a framework that obliges individual events to fit into certain trends but an approach that identifies moments that disrupt existing relations and create a contested space where change is possible. This does not necessarily mean that change is perceived as desirable or that the moment cannot be reinterpreted in such a way as to shut down its revisionary potential and consolidate existing patterns of ideological belief. Analysis centred on the dominant fiction is able to progress in this fashion because the dominant fiction is a dynamic system constantly in the process of creating itself. Also, it is at moments of trauma that the dominant fiction becomes visible, as it were, when the "reflecting surface" of ideology held up to society in order to work towards hegemony cracks and causes a breakdown in conventional mechanisms of soliciting ideological belief (SILVERMAN 1992, 24).

In terms of space, fragmentation undergoes significant changes as the arcs progress. Attacks on Avengers Mansion, specifically the way in which the incoming jet is framed, call to mind the coverage of the second plane to strike the WTC; seen by ZIZEK (2002, 15) as "the ultimate Hitchcockian blot, the anamorphic stain which denaturalised the idyllic well-known New York landscape," there are clear parallels between the framing of both jet crashes. Commentators speculated that a new world had been created in the wake of the 9/11 attacks but this was literally true in the Marvel universe – to avoid trauma, Scarlet Witch constructed an entirely new reality. In his own way, Iron Man attempts to do something similar by regulating and redistributing superheroes throughout the USA. As a response to the upheaval of *Disassembled* and *House of M*, to the disorder of *Civil War*, Iron Man seeks to impose order on both physical and ideological space. In its first real test, however, Iron Man's Initiative fails as the Skrulls seek to construct another reality on Earth – a mission that fails only due to a powerful restatement of an older world order.

Both in the way it is represented and in its effects, fragmentation is a force of significant dynamism in post 9/11 Marvel comic-books. Initially, fragmentation is represented most prominently through space, indicative of the way in which real-world spatial concerns had been disrupted by the terrorist attacks that forever altered the New York skyline. As the War on Terror shifted from a point of widespread unity to one of increasing division, fragmentation made itself felt at the level of narrative, mediating doubts about the place and the purpose of the US in the global order, something most clearly apparent in *Civil War*. *Secret Invasion* is at least in terms of narrative rather more straightforward; however, that there are continued meditations on fragmentation, and that the resolution offered is so clearly temporary, suggest that this is not engaging in explicitly conservative rhetoric about the need to safeguard the existing system indicates that, to some extent, fragmentation has been incorporated as an integral part of a coherent system – the prospect of a fully logical totality has been abandoned, or at least deferred, in time-honoured superhero fashion, until a later date.

In Zizekian terms (1999, 21), the presence of fragmentation within a supposed totality can be seen as "a particular element which subverts its own universal foundation," that is, an aspect which, although necessary for the construction of a particular social order, is also the point at which that order begins to break down, to fragment. Superheroes fit relatively smoothly into this framework: the societies in which superheroes reside are dependent on them for their continuation yet the presence of superheroes constantly subjects the same society to forces that threaten its extinction. Each of the arcs discussed here makes the same point in a slightly different way – a society founded on trauma includes within it the very elements that may ultimately lead to its collapse. When that society encounters another traumatic moment, when the workings of the dominant fiction are exposed, there is therefore a real threat of fragmentation. Most prominent in *Disassembled* and *House of M*, where a lack of organised resistance is apparent in a great deal of the narrative, fragmentation is explicitly externalised and visualised; *Civil War* and *Secret Invasion* works towards incorporating fragmentation as predominantly thematic rather than visual. Each arc ends with a reestablishment that is, to varying extents, a recalibration of what went before and in this respect the arcs mediate the functioning of the dominant fiction, drawing on new elements and dropping ones that have ceased to serve a purpose.

Superheroes are therefore a response to, as well as a guarantee of, fragmentation. Their exploits are directed towards the creation of a rational totality, a society free of superheroes, but this is endlessly deferred; in the meantime, the actions of superheroes explicitly identify points at which order has begun to breakdown, offering an attempt to work through points of tension in a form that can never offer what it promises – a society devoid of its symptom. Post 9/11, superheroes addressed real-world concerns in a way that acknowledged the contingency of both fictional and actual solutions, stressing the impossibility of short-term closure and acknowledging, through form and business model, the long-term persistence of the problems faced by US society.

ACKNOWLEDGEMENTS

I would like to thank Marvel for granting permission to reproduce the images in this article.

REFERENCES

BENDIS, B. M., J. CHEUNG et al. (2008): *New Avengers* #45. New York.

BENDIS, B. M., O. COIPEL et al. (2005): *House of M* #5. New York.

BENDIS, B. M., D. FINCH et al. (2004a): *The Avengers* #500. New York.
BENDIS, B. M., D. FINCH et al. (2004b): *The Avengers* #501. New York.
CARUTH, C. (1996): *Unclaimed Experience: Trauma, Narrative and History*. Maryland.
CHANG, N. (2002): *Silencing Political Dissent: How Post-September 11 Anti-Terrorism Measures Threaten our Civil Liberties*. New York.
DITTMER, J. (2005): Captain America's Empire: Reflections on Identity, Popular Culture, and Post 9/11 Geopolitics. *Annals of the Association of American Geographers* 95 (3), 626–643.
FALUDI, S. (2008): The Terror Dream: What 9/11 Revealed About America. London.
FREUD, S. (2001): *The Standard Edition of the Complete Psychological Works of Sigmund Freud: Beyond the Pleasure Principle, Group Psychology and Other Works*. London.
HUDLIN, R., S. EATON et al. (2006): *Black Panther* #18. New York.
JOHNSON, C. (2001): *Blowback: The Costs and Consequences of American Empire*. New York.
KAPLAN, E. A. (2001): Melodrama, Cinema and Trauma. *Screen* 42 (2), 201–205.
LAPLANCHE, J. and J. B. PONTALIS (1973): *The Language of Psychoanalysis*. London.
NEAL, A. (2005): *National Trauma and Collective Memory: Major Events in the American Century*. New York.
NICIEZA, F, S. JOHNSON et al. (2006): *Cable & Deadpool* #32. New York.
SILVERMAN, K. (1992): *Male Subjectivity at the Margins*. London.
WOLF-MEYER, M. (2003): The World Ozymandias Made: Utopias in the Superhero Comic and the Conservation of Difference. *Journal of Popular Culture* 36 (3), 497–517.
ZIZEK, S. (1999): *The Sublime Object of Ideology*. London.
ZIZEK, S. (2002): *Welcome to the Desert of the Real*. London.

Chapter 9 | *Marcus A. Doel*

AND SO. SOME COMIC THEORY COURTESY OF CHRIS WARE AND GILLES DELEUZE, AMONGST OTHERS. OR, AN EXPLICATION OF WHY COMICS IS NOT A SEQUENTIAL ART

Duration is defined less by succession than by coexistence.
(Gilles DELEUZE 1989, 60)

BUT

You must go on. I can't go on. I'll go on.
(Samuel BECKETT 2012)

In this chapter I advance a new understanding of comics that departs radically from the well-established characterization of comics as a 'sequential' medium. For the theorist of comics as a sequential medium, one moves through time as one moves through space: panel by panel. This pointillistic passage through space as if it were time strings together a time-line that has both duration (it goes 'on and on') and direction (typically from the past into the future via an insatiable present). As it progresses (on and on), the time-line comes to occupy the entire surface of the comics page, and swallows up the entire volume of a graphic novel. While there are all kinds of problems with this characterization of space and time, many of which I have laid bare elsewhere (DOEL 2004, 2010a), what I want to focus on in this chapter is the mischaracterization of comics as 'sequential' – irrespective of whether or not the sequence is uni-linear, multi-linear, non-linear, tabular, braided or aleatory. Everything will hinge on whether or not a panel is sutured to other panels to form a countable sequence that would move time on and on or whether it is sutured to the void.

The chapter begins with an analysis of the characterization of comics as 'sequential,' spatializing time by successively juxtaposing panels: either as a single linear sequence (AND THEN) or as several linear sequences (AND THEN, BUT ALSO). Much of the existing literature regards the later as a more sophisticated and potentially more critical form of comics than the former, since the interlacing or braiding of multiple linear sequences interrupts the seemingly smooth running of time, opening up the past, present, and future to an incessant reshuffling and interminable reworking. (Hence the affinity between comics and film.) However, this characterization of comics mistakenly regards time as if it were made up of points (panels) and lines (sequences). I therefore set about deconstructing this granularization and linearization of time, and argue that due to the absolute proximity of the void to every panel, panels can be neither enumerated nor sequenced. Rather than forming a whole, and moving on as a whole, panels remain still, plunging each and all into the void. The medium of comics presents a constellation of stills suspended in the void.

My argument is primarily advanced by recourse to the philosophy of Gilles DELEUZE (while drawing upon the concepts of Alain BADIOU, Jean BAUDRILLARD, Jacques DERRIDA, Michel FOUCAULT, and Jean-François LYOTARD along the way), and it is illustrated with some 'stirring stills' taken from the *œuvre* of Chris WARE, to purloin a fine phrase from Samuel BECKETT (2009). One of the things that I relish about WARE's comics is the fact that he peppers his work with words such as 'AND,' 'SO,' 'BUT,' 'MEANWHILE,' and 'THUS.' In this chapter I am attempting to move comics theory from the ill-conceived notion of 'sequencing' – no matter how radicalized – to the more fitting notion of 'voiding.' Just as one falls into the void of film via the cut, one falls into the void of comics via the gutter. But the void is not nothing, not least because of the absolute proximity of zero and infinity (SEIFE 2000). The void is inconsistent multiplicity, and everything that comes to exist and consist passes through it (BADIOU 2005a, 2009a, 2009b). Hereinafter, the gutter takes all.

AND SO

In *Understanding Comics*, MCCLOUD (1993, 9) offered a definition of comics, variations of which, and departures from which, have become staples of comics theory and scholarship ever since: "Juxtaposed pictorial and other images in deliberate sequence." The cumbersome phrase 'pictorial and other images' is intended to ensure that comics is open to all manner of images, loosely conceived, including words and pictures. Such images are not only set out *next* to one another (juxtaposed), they are set out to *follow* one another (sequenced).[1] They are arrayed and serialized, spatialized and temporalized, subjected to a two-fold regime of contiguity and succession. They are, in other words, ordered. Comics essentially repeats the following refrain: 'And so,' so to speak. Or better still: 'And so, on!' 'On' is the order-word that crystalizes the essence of comics. It is a word that is both spatial and temporal, reminding us that they are inseparable, since there is always already a 'becoming-space of time' and a 'becoming-time of space' (DERRIDA 1982). (Just think of the 'presence' of the 'present' or the phrase 'on time.') Wherever one finds oneself within the manifold pages of a comic, one must go on – one must pass on to what follows. I am reminded of what BADIOU says of

1 Since we are speaking of *following*, rather than say 'sequenced' I would have preferred to have said 'seduced' – in BAUDRILLARD's (1990) sense (*se-ducere*: lead away, lead astray), which he played off against 'producted' (*pro-ducere*: lead forward, make appear, force to appear) – but that would carry the chapter in a completely different direction (DOEL 2010b). Nevertheless, the text that follows is composed under the sign of seduction.

BECKETT's attempt to have done with language ('say on') in *Worstward Ho*: "There will be no termination of saying, no advent of the void as such. The *on* cannot be effaced" (BADIOU 2005b, 106). Even silence speaks volumes.

By going on and on relentlessly, it should come as no surprise to discover that MCCLOUD's definition of comics accords with BREDEHOFT's (2006, 870–871) claim that the medium's "architecture of narration" rests on "using space to represent time." Time is spatialized in comics. One moves through time as one moves through space: panel by panel, page by page, in an orderly and contiguous sequence which, for the Western reader, customarily runs from left to right, from top to bottom, and from front to back. However, MCCLOUD's definition may lead astray, especially if one fails to appreciate that it applies to the *form* of comics, rather than to the *content*, and that it only concerns the *ordering* of that form: on and on. It is a minimal definition, which boils down to the claim that comics is a 'sequential art' (EISNER 1992).

We can discern four paths that may lead astray. First, the 'on-going' may not be 'well' ordered. Juxtaposition may take place 'at a distance,' through folding, braiding, montage, resonance, differential repetition, et cetera. For example, GROENSTEEN's (2007) distinction between 'restricted' and 'general' arthology (dis/articulation, dis/jointure, dis/adjustment, et cetera) basically boils down to a distinction between *proximal juxtaposition* (the sequential or tabular connections that may be established between the panels on a page) and distal *juxtaposition* (similar connections that span different pages).[2] Second, the passage of content – the 'goings-on,' so to speak – need not be contiguous or successive. Obviously, the time that elapses between two panels may range from zero to infinity. For example, WARE often repeats identical panels (as if time *itself* stood still – not as an immobilized instant, but as an eternal duration), and occasionally juxtaposes the origin and the end of the world (which typically function as limit conditions of existence – Figure 9.1). Likewise with the span of space between two panels: one imperceptibly small step for Jimmy Corrigan; one giant leap across the universe for Rusty Brown. Third, the sequence need not be rectilinear or uni-directional ('Straight on'). Given an eight-by-eight grid of panels, for instance, there is no imperative for the Western reader to traverse it from left to right and from top to bottom. It could be traversed in reverse order. Or by a closed Knight's Tour.[3] Fi-

---

2 GROENSTEEN's distinction between restricted and general arthology echoes BATAILLE's distinction between restricted and general economy (BATAILLE 1991). However, the dialectical nature of the former is incompatible with the non-dialectical character of the latter (DERRIDA 1978). Arthology remains faithful to the production of sense; it cannot bring itself to truly risk losing sense. For while meaning and sense may be differed and deferred, they are nevertheless dutifully expected to take their proper place when they become due. In other words, arthology remains speculative rather than disseminative. It articulates the relationship between the restricted and the general in terms of a continuum (contraction and extension) rather than in terms of a dehiscence (deconstruction and de-consistence). This is why braiding leaves nothing shaken.

3 A knight's tour requires the knight to visit every square on a chessboard only once. If the tour ends one move away from the origin, then it is a *closed* tour, since the knight could repeat the

nally, what goes on and on need not be 'one and one' (DOEL 2001). There may be several sequences in play, such as a sequence of words and a sequence of pictures. In WARE's *Quimby the Mouse*, for example, "each lineation requires the reader to determine anew if the conjunctions linking a series of panels are those of movement through time or space, shifts in perspective or scale, or contiguities of mental or physical association. Temporal progress occurs in three and probably four directions and sometimes is indeterminate" (WIDISS 2010, 160). Meanwhile, WARE's graphic novel *Jimmy Corrigan, the Smartest Kid on Earth* not only interferes with the linearity of the narrative sequence, it also interferes with its speed. BANITA (2010) points to WARE's use of techniques of narrative deceleration and retardation, such as digression, defamiliarization, repetition, and excess (especially an excess of connectivity amongst seemingly disparate panels). "His techniques range from dividing a panoramic page into polyptychs that chart several different units of time ... to temporal overlaps and methods of stalling narrative progression" (BANITA 2010, 180).

And so, having enumerated the four-fold complication of what may take place when comics go 'on' we arrive at the essential formula that characterizes comics as a 'sequential' medium in which time is transmogrified into space: "[T]he comic strip is literally a map of time," says RAEBURN (2004, 11). Or again: "The diegetical horizon of each page [is] made up of what are essentially boxes of time" (CHUTE and DEKOVEN 2006, 769). Although it is far from a necessity, the spatialization of time in comics is often rendered in the form of a set of panels, each of which is framed by a border, separated from one another by gutters, and arranged on the page in tiers. Each panel is itself a constellation of images with its own duration (requiring explication, like a *tableau vivant*), each gutter registers a time and space lapse (whose own duration and extent are unbeknown to the reader), and the set of panels is sequenced in a well-ordered fashion from top left to bottom right. Obviously, these spatial-cum-temporal practices are not set in stone, and should be regarded as constraints that may be stretched, transgressed or broken. In FEAZELL's comic strip *The Incredible Mr Spot*, the eponymous Mr Spot literally purloins cash from his future self by dropping a fishing line into a lower tier of panels. "[T]he two-dimensionality of the page allows panels that represent different moments in time (according to the narrative line) to be vertically juxtaposed (raising the possibility that they are contiguous in time)," notes BREDEHOFT (2006, 873). "The fishing line, which crosses the boundary between panels, stands out as the very image of the way in which this joke operates by toying with conventional interpretive rules." Yet just because panels are contiguous in space does not necessarily mean that they are contiguous in time. Far from it – not only because of the time lapses and space lapses between panels, but also because of the four-fold complication of succession that I enumerated above.

---

same tour *ad infinitum*. PEREC (1996) famously structured his novel, *LIFE A USER'S MANUAL*, as a knight's tour of a 10-by-10 chessboard. Over the course of 99 chapters the reader takes a knight's tour of the 100 rooms and spaces of a Parisian apartment block at precisely a moment before eight o'clock in the evening on the 23rd of June 1975.

Panels can be expanded or contracted, frames cast off, borders breached, gutters occluded, and sequences short-circuited. KUHLMAN (2010) notes the affinity between WARE and Oubapo (*Ouvroir de bande dessinée potentielle*; Workshop of Potential Comics), since both undertake "formal experimentation in the medium and about the medium" (KUHLMAN 2010, 78). Since all forms of expression are produced under constraint (rules, conventions, norms, et cetera), WARE and Oubapo subject themselves to those constraints that "force the system out of its routine functioning, thereby compelling it to reveal its hidden resources" (BÉNABOU 1998, 41–42). For example, consider constraints that disrupt sequentiality, offering the would-be reader multiple entrances and exits that may short-circuit one another, lead into a labyrinth or give rise to lipogrammatic comics, tabular comics, palindromic comics or some other twisted form (Figure 9.2).

Moreover, we need to be attentive to the specificity and incommensurability of words and pictures, of reading and seeing (CHUTE 2006, DELEUZE 1988, HOLLAND et al. 2009). Comics "is a form that ... always refuses a problematic transparency, through an explicit awareness of its own surface" (CHUTE and DEKOVEN 2006, 767). Like MAGRITTE, WARE makes "pictures act like words [and] words act like pictures" (RAEBURN 2004, 19). And he has experimented with each of them separately as well as conjointly. For instance, he created "wordless strips – we might call them 'silent' strips" (RAEBURN 2004, 18) – in comics such as *Cat Daddy* and *Quimby the Mouse*. And just like the early days of 'silent' film, when films were still called 'comics,' these wordless strips never ceased to make sense through the histrionic gesticulation of a mute discourse. In strips that conjoin words and pictures, such as Figure 9.2, WARE exhibits the sight of words and the sense of pictures. Just as MAGRITTE hollowed out an ever-expanding "superficial abyss" (BAUDRILLARD 1988, 70) between a painted 'pipe' and a written 'pipe,' simply by stating the obvious – *This is not a pipe* – (FOUCAULT 2008), WARE hollows out a superficial abyss between seeing and reading in his strips. He has that rare ability to "make the word and the image equally legible and meet in the middle of the spectrum" (RAEBURN 2004, 19), not in order to establish a stable and harmonious equilibrium, but in order to set them in motion and agitation: each becoming other.

The experience of reading and seeing are different: "in principle the phenomenal text is not seen as such, the 'materiality of the signifier' vanishes in the interests of an immediate intellection of the signified" (BENNINGTON 1988, 56). Perversely, many want comics to become transparent, ostensibly to facilitate 'speed reading.' When comics tend towards pictograms, ideograms, and icons, CATES (2010, 96) claims that "their process of signification is diminishingly a matter of resembling the thing they represent, increasingly a matter of accepted symbolic conventions." CATES (2010, 96) then notes how WARE "describes a desire for his drawings to 'read like words' – for them to trigger meaning as immediately as a printed word will for a literate reader, 'so that when you see them you can't make yourself not read them.'" He wants his drawings to yield "immediate pictorial legibility," which would require "stylistic transparency" (CATES 2010, 96 and 97, respectively). Yet I would argue that in WARE's case, his desire is not for transparency and a speed-reading that would efface the texture of the text, so to speak, since he also exploits the reciprocal movement of letting his words be 'seen like pictures.' For the eye that 'reads' comics, "[t]he space of the text is barely spatial: it is *flat*, and the signs are linked according to a simple principle of horizontal contiguity" (BENNINGTON 1988, 57). For the eye that 'sees' comics, "the visual space of the word is complex, multi-dimensional and *deep*, oriented around the spatiality of my own body held within it" (BENNINGTON 1988, 57). WARE's comics demand a double reading: one flat and one deep.

Arguably, the only unshakable constraint is the one that declares that comics must *move on* – interminably. But even when this 'on-going' becomes impossible, one is compelled to move on. Comics take every conceivable abyss, asymptote, and impasse in their stride. "Try again. Fail again. Better again. Or better worse. Fail worse again. Still worse again" (BECKETT 2009, 81). And so it is with WARE's comics – try again, fail again, fail worse again (Figure 9.3).

In the wake of such lessening we come to appreciate that the attempt to frustrate the 'on-going' of comics as a sequential art requires an increasingly arduous 'moving worstward.' Even if time is arrested (reversed, even), space-as-time (which is to say, space in lieu of time) will inexorably move on. As SATTLER (2010, 216) astutely notes, "the language of comics lacks what we might call an imperfect tense, a way of depicting ongoing or habitual action in the past." It seems to be trapped in a *perpetual present*. Wherever one settles one's eyes, it always seems to be 'Now.' "The immediacy of comics continually jars the past tense into the medium's visual present" (SATTLER 2010, 217). Like film when it is subjugated to continuity editing and narrative space (HEATH 1981), comics *continuously re-centres* its narrative in the passage of the present, as if it were a window looking onto time itself: "the story's narrated past becomes experienced simply as a new narrative present" (SATTLER 2010, 216). This is exemplified by flashbacks: the past relived in its present, rather than in its past. WARE's *Building Stories* "confronts this problem directly, devising ways for the past and present to exist simultaneously, allowing the viewer to see the passage of time by maintaining all tenses at once ... [via] an artistic embodiment of the 'past imperfect'" (SATTLER 2010, 217). For instance, he keeps his images of the past firmly "lodged in the past" by allowing the present-day voice of an old woman to emerge from a great many of her younger selves. ... [This is] the present that gives the past its 'pastness'" (SATTLER 2010, 217). Yet even as the passing of time is pinned/penned down, so that the reader can experience remembering, space-as-time, space in lieu of time, nevertheless cannot not but move on. "[T]o continue to worsen becomes more and more difficult" (BADIOU 2005b, 107). One reaches not an impasse, but an exponential effort. Worsening is exhaustive rather than terminal. "So leastward on. So long as dim still. Dim undimmed. Or dimmed to dimmer still. To dimmost dim. Leastmost in dimmost dim. Unworsenable worst" (BECKETT 2009, 108). Thus, Figure 9.4.

And so, we are left teetering on the asymptotic cusp of the void with four possibilities. First, go on again – with the moving on of space-as-time, whether 'well' ordered or 'dis' ordered: linear, curvilinear, labyrinthine, broken, fragmented, multiplied, ramified, pullulated, superimposed, short-circuited, et cetera. Second, fail again – go worstward on or leastmost on amid the undimable dim, exhausting the combinatorial matrix. But rest assured that it will go on. Why? Because the combinations and permutations run to infinity. Third, stop. And wait. Stop and wait in suspended animation, like Bartleby the Scrivener. (This is how comics are given to us: stilled.) Finally, just let go. Let go of every line, starting with the life-line to which comics theory has clung: *sequence*.

MEANWHILE

Before moving on to consider the third and fourth possibilities – 'stoppage' and 'letting go' – it is worth lingering over the first possibility: going on. Why? Because this will signal our departure from McCloud's (1993) definition of comics, and its convolution by Groensteen (2007). McCloud uses his definition to granulate, linearize, and enclose time, and to distance comics from other sequential media, especially animation and film, all of which is extremely problematic.

Since time is spaced out on the surface of the page, it is decomposed into points, lines, surfaces, and volumes. As I have already noted, these are not points in time ('instants'), but durations of time ('meanwhiles'). Every panel *takes (its) time*. This may, of course, be a 'moment,' so-called – asymptotically tending towards a pure present (punctum), yet ineluctably bearing the trace of intervallic spacing and *différance* – but it is always a *period of time*: the more or the less. McCloud (1993, 95) gives the example of a single elongated panel within which a sequence of actions, reactions, and dialogue unfolds, and mentions that "in some respects this panel by itself actually fits our definition of comics! All it needs is a few gutters thrown in to clarify the sequence. ... One panel operating as several panels" (McCloud 1993, 97). There is not only an extensive sequence, then, *between* panels, but also an intensive sequence *within* panels, ready and waiting to unfurl: like a *tableau vivant*, diorama or panorama (Hecht 1993, Lant 1995). For once one embarks on the process of cutting up time, one will soon discover that it can be cut up indefinitely. Such is the decomposition of time into an infinite number of infinitesimal slivers whose duration and extent asymptotically approach zero. There is no stopping point. The cut never arrives at the limit (an indivisible), and always eludes the present (Deleuze 1990).

BUT

Given that every slice and morsel of spatialized time expresses *duration*, it is odd that McCloud (1993) should repeatedly insist that the medium of comics needs at least *two* panels. For although he acknowledges that single panels may be in the style of 'comic art' (i.e. they may draw upon the 'language,' 'vocabulary,' and 'grammar' of comics), he nevertheless insists that they are not part of the medium of comics (i.e. they do not have the requisite form). This is ostensibly because "there's no such thing as a sequence of one" (McCloud 1993, 20). However, McCloud is mistaken. Since every panel is accorded a portion, slice, sliver or trace of duration – if only infinitesimally and tending towards the limit of pure instantaneity – it 'contains' an elapse of time, and therefore a time-sequence that may be unfolded. Every panel *takes time*, just as it *takes place*. And since it endures, the time that it *takes* can be prolonged indefinitely. (Any panel whatsoever from Ware's œuvre can be lingered over interminably, can be made to stutter or stammer, and can be repeated or iterated ad infinitum.) This means that no two panels can ever be truly contiguous. The lapse takes all. (Such is the affinity between comics and the Lacanian unconscious.) What is at stake in the definition of comics as a 'sequential' art is neither contiguity nor continuity, but transitivity: if a relation applies between successive members of a sequence, it must also apply between any two members of the sequence taken in order. And as we have already seen, the 'relation' that concerns comics is space-as-time: moving on in time, going forward in time, et cetera. Obviously, transitivity need not hold (Law 2000). A trivial example: Alice grows taller. Evidently, she becomes taller than she was. (She grows up.) But she also becomes shorter than she will be. (She grows down.) Clearly, she isn't taller and shorter 'at the same time.' Only the *becoming* taller and *becoming* shorter coexist. Such is the "simultaneity of becoming whose characteristic is to elude the present... [B]ecoming does not tolerate the separation or the distinction of before and after, or of past and future. It pertains to the essence of becoming to move and pull in both directions at once: Alice does not grow without shrinking, and vice versa" (Deleuze 1990, 1).

Moreover, if we were to open up the play of form beyond 'pure time' to encompass the capacity of comics to 'show' and 'tell' through pictures and words, which is another staple of comics theory, then it would be immediately apparent that even a blind and mute panel – an 'imageless' or 'obscene' panel (Agamben 2002, and Baudrillard 1988, respectively) – can be explicated. Suffice to say that the differential repetition unleashed by the incommensurability of words and pictures hollows out representation into an infinite abyss while proliferating simulacra everywhere (Deleuze 1994, Doel 2010a, Lyotard 2011). Just look in awe at what Foucault (2008) drew from Magritte's painting, *The Treachery of Images / This is Not a Pipe* (1928–1929). Even an ostensibly 'empty' panel, which shows no-thing and says no-thing, *takes time*. However, there is a much more significant mistake in play in McCloud's (1993) exclusion of a solitary panel from the medium of comics. This is the mistake of treating sequencing as if it were counting. Even if we were rash enough to accept that a sequence needs to be countable – which I would caution against for reasons that will become apparent – then we would have to say that there is in fact a sequence of one. Here as elsewhere, one comes after zero (not to mention the negative numbers and an infinite set of fractions, as well as the reversion of the series, since the first may only take its place after the second has come to pass). Sudden thought: what if every panel, no matter how many of them have been arrayed in a series, came after zero? So that every one of them – the first, the second, the third, the fourth, et cetera – did not so much follow one another in a 'well ordered' sequence, but followed on from zero? If that were the case, then every panel would be in absolute proximity to the void. One might then begin to imagine that every panel is *drawn* from the void. (Isn't this what the blank page and the gutter gesture towards?) Each panel would then follow from the void, and come to rest *alongside* other panels through interpolation rather than succession. Having been *forced* to appear, every panel would be illicit and purloined: wholly incommensurable with the void, and completely alienated from its neighbours. Each panel would be indifferently detached. Hereinafter, every reading would be a forced reading.

Oblivious to the void, McCLOUD (1993) enumerates various forms of 'transition' between panels: such as moment-to-moment, action-to-action, subject-to-subject, and scene-to-scene transitions. The 'transition' forces the two panels to become commensurable (i.e. exchangeable), yielding what McCLOUD calls 'closure.' This is when the 'two' become 'one,' through integration, synthesis, fusion, assimilation, sublation or suchlike. "By creating a sequence with two or more images, we are endowing them with a single overriding identity, and forcing the viewer to consider them as a whole," he says (McCLOUD 1993, 73). Yet his account is blind to the most important 'transition' of all: the 'transition' from the void to the panel. And since there is *no common measure* between them (there is, strictly speaking, *nothing* to 'draw upon' or to 'say on'), "it is not a change, but a separation" (BADIOU 2005b, 119). There is, then, nothing 'between' the void and the series of panels, least of all a 'transition.' Comics must *force* some thing to appear (and disappear) by way of the void (q.v. footnote 1). In this context, it is interesting to consider the enduring image of the artist faced with a blank page or blank canvas, which WARE repeatedly evokes. In *The Acme Final Report to Shareholders and Saturday Afternoon Rainy Day Fun Book*, WARE (2005) re-envisions the history of art, stretching from cave painting into the far-flung future. Three large panels depict a pre-historic artist, a Renaissance artist, and an artist of the future, respectively. Each artist "is shown in the same pose ... facing the terror of the blank page ... suggesting the timeless anxiety of the blank page" (ROEDER 2010, 71–73. See Figure 9.5).

MAURO (2010, para. 24) offers this interpretation:

> Each of the artists ... looks absently into space before choosing what to draw. The implication Ware offers is to simply look out the window. ... The caveman need only turn to gaze out of the cave for images of life and death. The Renaissance man need only look out the window for images of a beautiful culture. The man of the future need only look outside to see the complexity and alienation of his society.

Nothing could be further from the truth. It is never a matter of 'just turning' and 'just looking.' On the one hand, this is because: "The painter does not paint on an empty canvas, and neither does the writer write on a blank page; but the page or canvas is already so covered with preexisting, preestablished clichés that it is first necessary to erase, to clean, to flatten, even to shred, so as to let in a breath of air from the chaos that brings us the vision" (DELEUZE and GUATTARI 1994, 204). Listen to what SATTLER (2010, 208) says about WARE's visual style, which "is defiantly *anti*-expressive, or what Ware calls 'flat,' 'harsh,' 'mechanical,' and 'banal.' This superficial coolness and purity of design, Ware insists, brings a measure of silence to his 'clattery' medium." On the other hand, in tending *towards* zero, one asymptotically *approaches* infinity: "Zero and infinity are two sides of the same coin," "eternally locked into a struggle to engulf all numbers," says SEIFE (2000, 131 and 145, respectively). As one moves worstward and leastmost on into the dim – in "the direction of the worst" (BADIOU 2005b, 97) – one hollows out an infinite abyss whose surface area tends towards infinity and whose volume tends towards zero, exactly like a Menger Sponge. Indeed, WARE's comics often resemble a Sierpinski carpet (Figure 9.6).[4]

And since "*Everything now returns to the surface*," as DELEUZE (1990, 7) once put it, this may be an appropriate juncture to consider MCCLOUD's attempt to distance comics from animation and film. For comics, animation, and film are all 'sequential media' that mobilize the 'surface of things' – screens, mirrors, paper, et cetera – as 'static vehicles' (VIRILIO 1994) for all manner of 'motionless trips' (DELEUZE and GUATTARI 1988). Along with a myriad of other pre-cinematic devices, including the camera obscura, panorama, diorama, stereoscope, and magic lantern, they have all sought to re-present, re-produce, and re-engineer space and time (CLARKE and DOEL 2005, 2006). To distance comics from film and animation, MCCLOUD (1993) emphasizes that in the medium of comics *still* images are juxtaposed, which thereby necessitates an active reader who forges sense and meaning through a productive engagement with the text. CHUTE and DEKOVEN (2006, 770) concur, arguing that:

> While seminal feminist criticism has detailed the problem of the passive female spectator following and merging helplessly with the objectifying gaze of the camera, the reader of graphic narratives is not trapped in the dark space of the cinema. She may be *situated* in space by means of the machinations of the comics page, but she is not ensnared in time; rather, she must slow down enough to make the connections between image and text and from panel to panel, thus working, at least in part, outside of the mystification of representation that film ... often produces.

On the basis of this simplistic mischaracterization of film, the reader of comics supposedly has more autonomy than a spectator of films, and some have suggested that WARE's comics are especially accommodating to the active reader by encouraging "a reading practice that may move freely backwards and forwards and across the page in numerous directions" (BENNETT and JACKSON 2009, para. 5). Similarly, HATFIELD (2005, 33) argues that "comic art is a form of *writing*," largely "as a reaction against the comparison of comics to cinema and other mechani-

4 To create a Sierpinski carpet, begin by dividing a square into a 3-by-3 grid of 9 sub-squares, and remove the central one. Repeat this procedure on each of the remaining 8 sub-squares, and recursively on all of the subsequent smaller and smaller sub-squares *ad infinitum*. The area of the carpet tends towards zero. To create a Menger sponge apply the same process to a cube: divide the cube into 27 sub-cubes, remove the central sub-cube on each face and the sub-cube at the centre of the cube. Repeat this procedure on the remaining 20 sub-cubes, and recursively on all subsequent smaller and smaller sub-cubes *ad infinitum*. The volume of the sponge tends towards zero while its area tends towards infinity.

cally paced, hence comparatively 'passive,' forms of visual communication." Like still images, writing calls for an active reader rather than a passive spectator; a reader who must forge meaning from a collection of stills. "I don't like to think of my work as 'cinematic,'" says WARE (2000, unpaginated). "A movie is *passive* – you're watching it, taking it in. With a comic strip, it's completely *active*: you have to read it, search it for meaning, for the connection with your entire experience and your memory. Yes, you do have the illusion of watching something happen in a comic strip – but if it's done well, it comes alive on the page like a novel." By constructing his comics as puzzles, KUHLMAN (2010, 86) argues that WARE produces what BARTHES called "a 'writerly' text, as opposed to a 'readerly' one; the former demands an active reader who 'produces' the text, while the latter only reinforces passivity and consumption." The comics reader, then, is put to work twice over: once in terms of the effort that it takes to bring the inert panels to life; and again in terms of the effort that it takes to solve the puzzles. However, I would prefer not to regard WARE's comics as puzzles, but as problematics. WARE never offers his readers a final solution to the 'problems' that he articulates and unfolds. There is, strictly speaking, nothing to solve, and nothing to explain. Every denouement is invariably belated and bathetic.

MCCLOUD also tries to distance comics from animation and film by claiming that they mirror one another. Comics are characterized by still images, spatialized time, and active subjects, while films are characterized by animated images, temporalized space, and passive subjects. However, not only is each of these oppositions that is tasked with keeping comics and film poles apart susceptible to deconstruction, but each of them is rendered problematic by the historical development of the media in question. Since I have elsewhere written extensively on the stillness and animation of the cinematic image, as well as its spatialization of time and temporalization of space, and its subjectivation of active and passive subjects (CLARKE and DOEL 2005, DOEL and CLARKE 2007), including an essay within this 'Media Geography at Mainz' series (DOEL 2008), I do not intend to labour the point. Suffice to say that the spatialization of time and temporalization of space within comics and film are of a piece. Attempts to distance comics from film invariably fall back on caricatures of both media. Rather than distance one medium from the other, or disparage one medium or the other, it would be preferable to consider their common fate – which, as we have begun to see, concerns their relationship with the void.

When all is said and done, MCCLOUD's suturing of his definition of comics to the operation of closure is an attempt to form a 'smooth space' out of the 'striated medium' of comics. "Comics panels fracture both time and space, offering a jagged, staccato rhythm of unconnected moments. But closure allows us to connect these moments and mentally construct a continuous, unified reality ... Comics is closure" (MCCLOUD 1993, 67). This closure of juxtaposed panels is meant to *pin down* time, and then, having anchored time into place, the sequence will also have been *pinned down,* and therefore *set* in motion. GILMORE (2010, 147) objects that MCCLOUD's "notion of closure, which typically operates on a panel-to-panel level, doesn't explain larger-scale graphic narrative structures," and advances

GROENSTEEN's structuralist notion of 'braiding' instead: "braiding deploys itself simultaneously in two dimensions, requiring them to collaborate with each other: synchronically, that of the co-presence of panels on the surface of the same page; and diachronically, that of the reading, which recognizes in each new term of a series a recollection or an echo of an anterior term" (GROENSTEEN 2007, 147). GROENSTEEN refers to the synchronic dimension as the 'plane of the system' and the diachronic dimension as the 'plane of the process,' which are 'braided' together whilst reading.

> By its nature, a story develops in length in a linear and irreversible manner ... reinforced by ... the 'bookish form' of the comic. ... With respect to comics, this disposition finds itself constantly embattled, and in a certain measure neutralized, by the properties [of] the panels. The network that they form is certainly an oriented network, since it is crossed by the instance of the story, but it also exists in a dechronological mode, that of the collection, of the panoptical spread and of coexistence, considering the possibility of translinear relations and plurivectoral courses. ... [R]eading in this case is the operation that puts into tension the *plane of the process* and the *plane of the system*. The panels of the disseminated series only form a constellation to the degree that the reading detects and decrypts their complementarity and interdependence (GROENSTEEN 2007, 147).

Through the 'braiding' of the plane of the system and the plane of the process, everything will have been knotted down – or 'quilted,' to borrow a motif from Lacan – via a loopy enchainment of images in spatialized time. No wonder BERGSON balked at the notion of spatialized time (BERGSON 1999, DELEUZE 1991). A sequence of immobilized sections – whether photographic stills, cinematic frames or comics panels – will never be able to regain lost time. DELEUZE (1986, 1) phrased it superbly: "On the one hand, you can bring two instants or two positions together to infinity; but movement will always occur in the interval between the two, in other words behind your back. On the other hand, however much you divide and subdivide time, movement will always occur in a concrete duration; thus each movement will have its own qualitative duration."

NOW

Much critical effort has been expended striving to demonstrate how comics might depart from linear time. And yet, if time cannot be pinned down and fixed into place, if time eludes its spatialization and enclosure, then it cannot be kept in line. In this regard, comics theory has much to learn from the way in which editing enabled early film (ca. 1903–1910) to express what CHRISTIE (1994, 33) aptly calls "vernacular relativity," conveying the "mechanization, jerkiness, and rush of modern times," and heightening the "public consciousness of differential speed" (KERN 1983, 117 and 130, respectively). Film also expanded "the sense of the present" by allowing filmmakers "to splice open a moment and insert a number of simultaneous activities" (KERN 1983, 70 and 71, respectively). In so doing film become a critical, explosive, and potentially revolutionary medium that would need to be muffled by the formation of continuity editing and narrative space.

Film enabled the fleeting moment to be extracted from its quotidian flow, so that it could become a material to be engineered: stopped, expanded, diminished, reversed, accelerated, decelerated, cut, spliced, transformed, et cetera (BURCH 1990, ELSAESSER and BARKER 1990). Like film, comics have the capacity to re-engineer space and time, to express 'vernacular relativity,' and to explode "the homogeneity of the epoch, interspersing it with ruins – that is, with the present" (BENJAMIN 1999, 474). And like film, these capacities have been subjected to a myriad of attempts to muffle and disable them, including developments akin to the advent of continuity editing, the formation of narrative space, and the continual re-centring of the viewer's point of view. Invariably, however, the more that scholars have struggled to resist the muffling of comics' explosive potential, the more they tend to tie themselves down. You can multiply the number of lines all you want, but it will not alter the fact that the text is being tied down: all the more so, in fact. Sense cannot be pinned down or tied into place (DELEUZE 1990, DERRIDA 1981).

Here as elsewhere, 'space' poses a problem. In comics theory, the problem is posed in terms of 'sequence,' 'line,' and 'dimension.' This is how BREDEHOFT (2006, 872) articulates the problem:

> In the architectures of language and cinema ... the narrative line ... involves the crucial element of time-sequencing. ... The nature of time-sequencing itself, however, whether in chronology or narrative line, is that time is uni-directional and irreversible: time passes, or we pass through time. The underlying metaphor, of course, is that time functions like one dimension of space: the narrative line is linear, precisely because language itself (or the procession of images that we see in film) is sequenced uni-directionally in time. In contrast to film or language-based narration, however, the medium of comics offers the possibility of a narrative mode that disrupts time-sequencing itself, and it appears to be the case that it is the specifically two-dimensional architecture of the comics page that allows comics narration to break the linearity of a time-sequenced narrative line.

Leaving to one side the reductive characterization of language and film, I simply want to note that for BREDEHOFT it is not so much the comics page that is two dimensional, but the multi-paneled page. "[T]he two-dimensionality of the page allows panels that represent different moments in time (according to the narrative line) to be ... juxtaposed (raising the possibility that they are contiguous in time)" (BREDEHOFT 2006, 873). (And so on this account, the split-screen would herald the cinematic screen's becoming-two-dimensional, and film's becoming-comics.) And short-circuiting time sequences also short-circuits causal sequences. For example, in *Jimmy Corrigan, the Smartest Kid on Earth*, WARE includes pages that present multiple sequences that cross-cut one another: "to read this page effectively does not depend on choosing a correct sequence for the juxtaposed images or a correct sequence of narrative lines, but to recognize that each individual sequence and line is insufficient to the narrative purposes of the page as a whole" (BREDEHOFT 2006, 878). However, despite the fact that "the two-dimensional architecture of the comics page readily accommodates itself to being read or interpreted as embodying or employing one, two, or even multiple simultaneous narrative lines," BREDEHOFT (2006, 884) claims that "Our perception of the comics page, apparently, remains primarily linear or multilinear." Opening up two dimensions, then, is insufficient to 'break' the hold of the line. Everything continues to move – *on*. Such is spatialized time's relentless melt.

So, WARE makes another attempt to 'break' the hold of the line by including templates for three-dimensional paper models within his comics, which the reader is encouraged to make in 'real life.' For instance, in *The Acme Final Report to Shareholders and Saturday Afternoon Rainy Day Fun Book* (2005), WARE proffers: "Your very own miniature working 'Acme novelty library' to cut-out, fold and assemble," which includes a little cabinet and tiny comics. In *Jimmy Corrigan, the Smartest Kid on Earth*, WARE includes a model of a zoetrope, which one of the characters makes elsewhere in the book. More frequently, WARE includes diagrams of the narrative amid the narrative. These diagrams usually explicate contextual and conjunctural elements of the narrative, such as an elliptical tour of the milieu within which the narrative circulates or a potted family history. For example, Figure 9.7.

BREDEHOFT (2006, 888) notes that PRAGER (2003) characterizes these "quasi-mathematical diagrams and flow charts" as 'map strips' that function like Mobius strips, "since they ostensibly have neither beginning nor end." BREDEHOFT (2006, 888) demurs from this characterization, arguing that they "have a purposeful, rather than merely unending, circularity." Meanwhile, CATES (2010, 94) offers another interpretation: "By presenting an alternative to the steady sequential march of comics narrative, Ware's diagrams allow the sort of retrospective or omniscient view of the story's events that other authors might reserve for captions in the voice of a third-person narrative." Yet the narrative that is diagrammed is invariably of a different order and magnitude to the narrative into which it is inserted, and more often than not the 'perspective,' 'information,' and 'purpose' that they serve are of no consequence. They are superfluous.

Earlier in the chapter I suggested that comics could break every conceivable constraint, except for the imperative to '*move on*,' and we have repeatedly seen that this is the case. According to CHUTE and DEKOVEN (2006, 780), "the import and meaning of graphic narrative is in how images interact with text, and in how they interact with other images on the page, moving time forward spatially." And since there are always (at least) two narrative tracks – the verbal (which may be mute) and the visual (which may be blind), each of which may be ramified to infinity and beyond – there are multiple narrative temporalities moving forward spatially.

> In comics, the images are not illustrative of the text, but comprise a separate narrative thread that moves forward in time in a different way than the prose text, which also moves the reader forward in time. The medium of comics is cross-discursive because it is composed of verbal and visual narratives that do not simply blend together, creating a unified whole, but rather remain distinct (CHUTE and DEKOVEN 2006, 769).

WARE's diagrams are significant because they 'break' this seemingly foundational constraint: that the medium 'must' *move on*. First, they interrupt the flow of transitions. They cut the narrative dead. Second, they demonstrate the redundancy of a supplementary dimension, a dimension within which the array of panels (either in whole or in part) would *make sense* (either finally or partially). Hereinafter, "[t]here is nothing to explain, nothing to understand, nothing to interpret" (DELEUZE 1977, 114). This casts WARE's fondness for 'ordinariness' in a new light. According to FINK BERMAN (2010, 195–197):

> Ware's version of 'the ordinary' is a living moment-by-moment, dwelling on the micro-gestures that narratives usually elide; it is the bored killing of time. ... What distinguishes Ware's storytelling style from that of most Western comics is the predominance of moment-to-moment transitions, in which the change and the lapse of time between panels is minimal. Ware's action-to-action transitions also function as moment-to-moment transitions, as he primarily records micro-actions.

Similarly, BANITA (2010, 178) argues that WARE wants to mobilize "the obsessively uncomfortable slower-than-real time in which the *Jimmy Corrigan* narrative plays out" to recalibrate "what we consider to be the speed of comics as a medium" – not only to acknowledge the time-consuming labour that is invested in their crafting but also to restore a more fitting 'speed' for reading comics. (One can see a 'dignified labour' and 'slow comics' movement beginning to stir.) On this account, WARE's comics stand opposed to the acceleration of comics production and comics consumption. BANITA (2010, 177) even goes so far as to suggest that *Jimmy Corrigan*'s painfully slow pace is pitched against modernity's acceleration and standardization of time, a suggestion that chimes with WORDEN's (2010, 110) claim that "Ware's work is best viewed as a catalog of modernism's ruins, an archive that illuminates neglected, lost, and forgotten possibilities."

By introducing an immanent break within the machinations of spatialized time, each of WARE's diagrams "is an event creating an *afar*. It is an incalculable distancing" (BADIOU 2005b, 119). But so too with each 'ordinary' panel. A speedy reading and a slow reading both depend upon a determinable and calculable distance between panels, otherwise there would be no passage and transition. But here we experience an incalculable distance 'between' panels, which not only means that the 'distance' cannot be crossed, but also that the 'next' position is undecidable. In other words, there are no points – least of all anchor points – to pin things down and emplot a series of lines. Given any panel whatsoever, we always find ourselves in the middle of things (DELEUZE and GUATTARI 1986).

Earlier in the chapter I noted that insofar as it follows zero, 'one' is always already sequential, which also means that every sequence – whatever its quantitative span – remains in touch with the void: not only at the origin, but at every point along the sequence. By way of the void "a barely perceptible displacement disjoints all the articulations and penetrates all the points welded together," says DERRIDA (1978, 260). "A trembling spreads out which then makes the entire old shell crack." The sequence of panels is not so much held together by a multitude of lines, which can be endlessly interlaced, braided, and knotted, but halved together and cracked open by the void, which, like 'difference' for DELEUZE (1994) and '*différance*' for DERRIDA (1978), suffuses all and takes all. Here as elsewhere, what takes place in comics is the void. In film, one touches the void via the cut. In comics, one touches the void via the gutter.

Ironically, McCloud was on the right track when he defined comics as sequential, but seems to have been derailed by mistaking a sequence (an order) for a count (one and one). "The panel acts as a sort of general indicator that time or space is being divided" (McCloud 1993, 99). Space and time are divided, sectioned, and apportioned. But, he continues: "The durations of that time and the dimensions of that space are defined more by the contents of the panel than by the panel itself" (McCloud 1993, 99). And so we arrive at the pivotal formula for comics theory: *some* time, *some* space. Neither *one* time, nor *one* space. Strictly speaking, "[t]he *one* as such ... is not. It is always the result of a count, the effect of a structure" (Badiou 2005a, 90). Every 'one' that goes 'on and on' only does so in so far as it is *forced* so to do.[5] In fact, "[t]he thinking of number obliges thought to *return* to zero, to the infinite and ... a total dissipation of the One" (Badiou 2008, 14–15). All that we are left with is a cutting up of the void; its sectioning and apportionment. And there is nothing to link these cut-outs and fold-outs, save the void.[6] There is no subterranean foundation, structure or order, nor is there any higher synthesis, unity or totality, within which the set of sections and array of panels would finally – or even provisionally – come together to make sense. We must bid farewell to the dream that "comics rely on dialectical relationships between the fragment and the whole; each panel is both discrete and bound to its predecessors and antecedents" (Worden 2010, 108). There is only ever *some* comics – the more or the less. *Some* more. *Some* less. That's all.

> We live today in the age of partial objects, bricks that have been shattered to bits, and leftovers. We no longer believe in the myth of the existence of fragments that, like pieces of an antique statue, are merely waiting for the last one to be turned up, so that they may all be glued back together to create a unity that is precisely the same as the original unity. We no longer believe in a primordial totality that once existed, or in a final totality that awaits us at some future date. ... We believe only in totalities that are peripheral. And if we discover such a totality alongside various separate parts, it is a whole *of* these particular parts but does not totalize them; it is a unity *of* all of these particular parts but does not unify them; rather, it is added to them as a new part fabricated separately (Deleuze and Guattari 1984, 42).

The 'One' and its 'Multiples,' which could ostensibly have been *set* in sequence and *taken up* 'as a whole,' give way to the 'Open' and its 'Multiplicities.' There is neither *one* comic nor *many* comics, just '*some comic*,' so to speak: *some* panels,

---

5 At this juncture we could consider comics from the perspective of nomadology and dromology, recasting comics theory in terms of great migrations and violent upheavals, deterritorialization and reterritorialization, voyaging smoothly and in striation, et cetera (Deleuze and Guattari 1988, Virilio 1986).

6 Since the linking of panels must go by way of the void, the distance between any two panels is incalculable and undecidable (ranging from zero to infinity), and the distinction between proximal and distal no longer holds. Near and far are neither given in advance nor settled by compositional and structural articulation. Near and far must be *forced* to appear. What I have in mind is rendered beautifully in Bioy Casares's (2003) *The Invention of Morel.* Or, more prosaically, simply look into the eyes of the one who is utterly indifferent to being your beloved. Or just stare into the eyes that look back at you in a mirror.

some gutters, *some* pages, *some* splashes, et cetera that will never have formed *a* Whole.

> The multiple *must be made*, not by always adding a higher dimension, but rather in the simplest of ways, by dint of sobriety, with the number of dimensions one already has available – always $n - 1$ (the only way the one belongs to the multiple: always subtracted). Subtract the unique from the multiplicity to be constituted; write at $n - 1$ dimensions. A system of this kind could be called a rhizome (DELEUZE and GUATTARI 1988, 6).

AND SO, FINALLY

> [D]uration divides up and does so constantly: That is why it is a *multiplicity*. But it does not divide up without changing in kind, it changes in kind in the process of dividing up: This is why it is a nonnumerical multiplicity, where we can speak of 'indivisibles' at each stage of the division. There is *other* without there being *several*" (DELEUZE 1991, 42).

WARE's *Building Stories* is very precisely *some* comic in the sense articulated above. The box in which it comes informs the would-be reader that it contains: "Everything you need to read the graphic novel *Building Stories*: 14 distinctively discrete books, booklets, magazines, newspapers, and pamphlets." The box also includes a "pictographic listing of all 14 items (260 pages in total)," including the box itself, most of which are tagged to a three-dimensional schematic of the ground floor of a house, although a few are placed in the hands and handbag of one of the characters included in the reading material.

On opening the box, the would-be reader finds a package of 13 items, and may notice that on three edges of the box there are a few disjoined and humdrum panels, one of which is located just around the corner from the blank edge. Some of the 'discrete' comics form a sequence, of sorts, such as when the report of Branford the Bee's perilous entrapment behind a pane of glass on page 3 of *The Daily Bee* newspaper is resolved on the fourth panel of a double-sided octaptych (Figure 9.8). Mostly, however, there are just passing encounters, and if one *forces* them to form a sequence, they singularly fail to add up.

THUS

> Not to maintain together the disparate, but to put ourselves there where the disparate itself *holds together*, without wounding the dis-jointure, the dispersion, or the difference, without effacing the heterogeneity of the other (DERRIDA 1994, 29).

With *Building Stories*, we may finally return comics theory whence it came: *some* comic, *some* panel, and *some* ink, all of which fall upon the page like rain or snow (DOEL and CLARKE 2009; Figure 9.9).

MEANWHILE

Everything takes (its) place in the gutter. As WILDE and NIETZSCHE may have said: we are all in the gutter, but some of us are looking into the void – and the void is smiling back, like *Sparky the Cat*.

ACKNOWLEDGEMENTS

I am extremely grateful to Chris Ware, who very kindly allowed his work to be reproduced in this chapter. I am also grateful to David Clarke, Jason Dittmer, and Richard Smith for their comments.

REFERENCES

AGAMBEN, G. (2002): Difference and repetition: on Guy Debord's films. T. MCDONOUGH (Ed.): *Guy Debord and the Situationist International: Texts and Documents*. Cambridge, 313–320.

BADIOU, A. (2005a): *Being and Event*. London.

BADIOU, A. (2005b): *Handbook of Inaesthetics*. Stanford.

BADIOU, A. (2008): *Number and Numbers*. Cambridge.

BADIOU, A. (2009a): *Theory of the Subject*. London.

BADIOU, A. (2009b): *Logics of Worlds*. London.

BANITA, G. (2010): Chris Ware and the pursuit of slowness. M. BALL and M. B. KUHLMAN (Eds.): *The Comics of Chris Ware: Drawing is a Way of Thinking*. Jackson, 177–190.

BATAILLE, G. (1991): *The Accursed Share: Volume 1*. New York.

BAUDRILLARD, J. (1988): *The Ecstasy of Communication*. New York.

BAUDRILLARD, J. (1990): *Seduction*. London.

BECKETT, S. (2009): *Company, Ill Seen Ill Said, Worstward Ho, Stirrings Still*. London.

BECKETT, S. (2012): *Trilogy: Molloy, Malone Dies, The Unnamable*. London.

BÉNABOU, M. (1998): Rule and constraint. W. MOTTE (Ed.): *Oulipo: A Primer of Potential Literature*. London, 40–47.

BENJAMIN, W. (1999): *The Arcades Project*. Cambridge.

BENNETT, J. and C. JACKSON (2009): Graphic whiteness and the lessons of Chris Ware's *Jimmy Corrigan*. *ImageTexT* 5(1). http://www.english.ufl.edu/imagetext/archives/v5_1/bennett-jackson/ (accessed September 1, 2013).

BENNINGTON, G. (1988): *Lyotard: Writing the Event*. Manchester.

BERGSON, H. (1999): *Duration and Simultaneity: Bergson and the Einsteinian Universe*. Manchester.

BIOY CASARES, A. (2003): *The Invention of Morel*. New York.

BREDEHOFT, T. A. (2006): Comics architecture, multidimensionality, and time: Chris Ware's *Jimmy Corrigan, the Smartest Kid on Earth*. *Modern Fiction Studies* 52 (4), 869–890.

BURCH, N. (1990): *Life to those Shadows*. London.

CATES, I. (2010): Comics and the grammar of diagrams. M. BALL and M. B. KUHLMAN (Eds.): *The Comics of Chris Ware: Drawing is a Way of Thinking*. Jackson, 90–104.

CHRISTIE, I. (1994): *The Last Machine: Early Cinema and the Birth of the Modern World*. London.

CHUTE, H. L. (2006): Decoding comics. *Modern Fiction Studies* 52 (4), 1014–1027.
CHUTE, H. L. and M. DEKOVEN (2006): Introduction: graphic narrative. *Modern Fiction Studies* 52 (4), 767–782.
CLARKE, D. B. and M. A. DOEL (2005): Engineering space and time: moving pictures and motionless trips. *Journal of Historical Geography* 31, 41–60.
CLARKE, D. B. and M. A. DOEL (2006): From flatland to vernacular relativity: the genesis of early English screenscapes. M. LEFEBVRE (Ed.): *Landscape and Film*. New York, 213–243.
DELEUZE, G. (1977): I have nothing to admit. *Semiotext(e)* 2 (3), 111–116.
DELEUZE, G. (1986): *Cinema 1: The Movement-Image*. London.
DELEUZE, G. (1988): *Foucault*. Minneapolis.
DELEUZE, G. (1989): *Cinema 2: The Time-image*. London.
DELEUZE, G. (1990): *The Logic of Sense*. London.
DELEUZE, G. (1991): *Bergsonism*. London.
DELEUZE, G. (1994): *Difference and Repetition*. London.
DELEUZE, G. and F. GUATTARI (1984): *Anti-Oedipus: Capitalism and Schizophrenia*. London.
DELEUZE, G. and F. GUATTARI (1986): *Kafka: Toward a Minor Literature*. Minneapolis.
DELEUZE, G. and F. GUATTARI (1988): *A Thousand Plateaus: Capitalism and Schizophrenia*. London.
DELEUZE, G. and F. GUATTARI (1994): *What is Philosophy?* New York.
DERRIDA, J. (1978): *Writing and Difference*. Chicago.
DERRIDA, J. (1981): *Dissemination*. Chicago.
DOEL, M. A. (2001): 1a. Qualified quantitative geography. *Environment and Planning D: Society and Space* 19 (5), 555–572.
DOEL, M. A. (2004): Poststructuralist geographies: the essential selection. P. CLOKE, P. CRANG and M. GOODWIN (Eds.): *Envisioning Human Geographies*. London, 146–171.
DOEL, M. A. (2008): From animated photography to film: the formation of vernacular relativity in early English films (1895–1908). C. LUKINBEAL and S. ZIMMERMANN (Eds.): *The Geography of Cinema – A Cinematic World?* Stuttgart, 87–99.
DOEL, M. A. (2010a): Representation and difference. B: ANDERSON and P. HARRISON (Eds.): *Taking-Place: Non-Representational Theories and Geography*. Aldershot, 117–130.
DOEL, M. A. (2010b): Production, Seduction. R. G. SMITH (Ed.): *The Baudrillard Dictionary*. Edinburgh, 171–174, and 186–189.
DOEL, M. A. and D. B. CLARKE (2007): Afterimages. *Environment and Planning D: Society and Space* 25, 890–910.
DOEL, M. A. and D. B. CLARKE (2009): The artistry of cities: Chris Ware's comic strips. T. BEYES, S.-T. KREMPL, and A. DEUFLHARD (Eds.): *ParCITYpate: Art and Urban Space*. Zurich, unpaginated.
EISNER, W. (1992): *Comics and Sequential Art: Principles and Practices from the Legendary Cartoonist*. Tamarac.
ELSAESSER, T. and A. BARKER (Eds.): (1990): *Early Cinema: Space, Frame, Narrative*. London.
FINK BERMAN, M. (2010): Imagining an idiosyncratic belonging: representing disability in Chris Ware's 'Building Stories.' M. BALL and M. B. KUHLMAN (Eds.): *The Comics of Chris Ware: Drawing is a Way of Thinking*. Jackson, 191–205.
FOUCAULT, M. (2008): *This is Not a Pipe*. London.
GILMORE, S. (2010): Public and private histories in Chris Ware's *Jimmy Corrigan*. BALL, M. and M. B. KUHLMAN (Eds.): *The Comics of Chris Ware: Drawing is a Way of Thinking*. Jackson, 146–158.
GROENSTEEN, T. (2007): *The System of Comics*. Jackson.
HATFIELD, C. (2005): *Alternative Comics: An Emerging Literature*. Jackson.
HEATH, S. (1981): *Questions of Cinema*. Bloomington.
HECHT, H. (Ed.) (1993): *Pre-Cinema History: An Encyclopaedia and Annotated Bibliography of the Moving Image Before 1896*. London.

HOLLAND, E. W., D. W. SMITH and C. J. STIVALE (Eds.) (2009): *Gilles Deleuze: Image and Text*. London.

KERN, S. (1983): *The Culture of Time and Space: 1880–1918*. Cambridge.

KUHLMAN, M. B. (2010): In the comics workshop: Chris Ware and the Oubapo. M. BALL and M. B. KUHLMAN (Eds.): *The Comics of Chris Ware: Drawing is a Way of Thinking*. Jackson, 78–89.

LANT, A. (1995): Haptical cinema. *October* 74, 45–73

LAW, J. (2000): Transitivities. *Environment and Planning D: Society and Space*. 18, 133–148

LYOTARD, J-F. (2011): *Discourse, Figure*. Minneapolis.

MAURO, A. (2010): "Mosaic thresholds": manifesting the collection and production of comics in the works of Chris Ware. *ImageTexT* 5(1). http://www.english.ufl.edu/imagetext/archives/v5_1/mauro/ (accessed September 1, 2013).

MCCLOUD, S. (1993): *Understanding Comics: The Invisible Art*. New York.

OLSSON, G. (2007): *Abysmal: A Critique of Cartographical Reason*. Chicago.

PEREC, G. (1996): *Life A User's Manual*. London.

PRAGER, B. (2003): Modernism in the contemporary graphic novel: Chris Ware and the age of mechanical reproduction. *International Journal of Comic Art* 5 (1), 195–213.

RAEBURN, D. (2004): *Chris Ware*. New Haven.

ROEDER, K. (2010): Chris Ware and the burden of art history. M. BALL and M. B. KUHLMAN (Eds.): *The Comics of Chris Ware: Drawing is a Way of Thinking*. Jackson, 65–77.

SATTLER, P. R. (2010): Past imperfect: 'Building Stories' and the art of memory. M. BALL and M. B. KUHLMAN (Eds.): *The Comics of Chris Ware: Drawing is a Way of Thinking*. Jackson, 206–222.

SEIFE, C. (2000): *Zero: The Biography of a Dangerous Idea*. London.

VIRILIO, P. (1986): *Speed and Politics: An Essay on Dromology*. New York.

VIRILIO, P. (1994): *The Vision Machine*. London.

WARE, C. (2000): Transcript of an interview with Chris Ware. *CNN.com*, October 3, 2000. http://edition.cnn.com/2000/books/news/10/03/chris.ware.qanda/ (accessed September 1, 2013).

WARE, C. (2003a): *Jimmy Corrigan, the Smartest Kid on Earth*. London.

WARE, C. (2003b): *Quimby the Mouse, or, Comic Strips, 1990–1991*. London.

WARE, C. (2005): *The Acme Novelty Library Final Report to Shareholders and Saturday Afternoon Rainy Day Fun Book*. London.

WARE, C. (2007a): *The Acme Novelty Library Number 18*. Montreal.

WARE, C. (2008): *The Acme Novelty Library Number 19*. Oak Park.

WARE, C. (2010): *The Acme Novelty Library Number 20*. Montreal.

WARE, C. (2012): *Building Stories*. London.

WIDISS, B. (2010): Autobiography with two heads: *Quimby the Mouse*. M. BALL and M. B. KUHLMAN (Eds.): *The Comics of Chris Ware: Drawing is a Way of Thinking*. Jackson, 159–173.

WORDEN, D. (2010): On modernism's ruins: the architecture of 'Building Stories' and *Lost Buildings*. M. BALL and M. B. KUHLMAN (Eds.): *The Comics of Chris Ware: Drawing is a Way of Thinking*. Jackson, 107–120.

APPENDIX OF FIGURES

Figure 1.1: First Frame of Dan Dare, in Eagle, 14/4/1950. Reproduced with the kind permission of the Dan Dare Corporation Limited.

Figure 1.2: The Windscale nuclear complex in 1957, with piles 1 and 2. Image supplied courtesy of the Nuclear Decommissioning Authority.

Figure 1.3: Opening frame of Dan Dare, in Eagle, 10/8/51. Reproduced with the kind permission of the Dan Dare Corporation Limited.

Figure 1.4: Frank Newbould poster. © IWM

Figure 2.1 (left and right): From 'I: The Gate of Tears', Snakes and Ladders.

"Our stories, our back-continuities that herded us down separate tracks toward toward this room, this mutual dark."

Figure 2.2 (left and right): From 'III Baghdad', Snakes and Ladders.

"Arabian nightmares smoulder in the corner of Red Lion Square."

Figure 2.3 (left and right): From 'III Baghdad', Snakes and Ladders.

"Some people pass their lives among its phosphorescent streetfairs..."

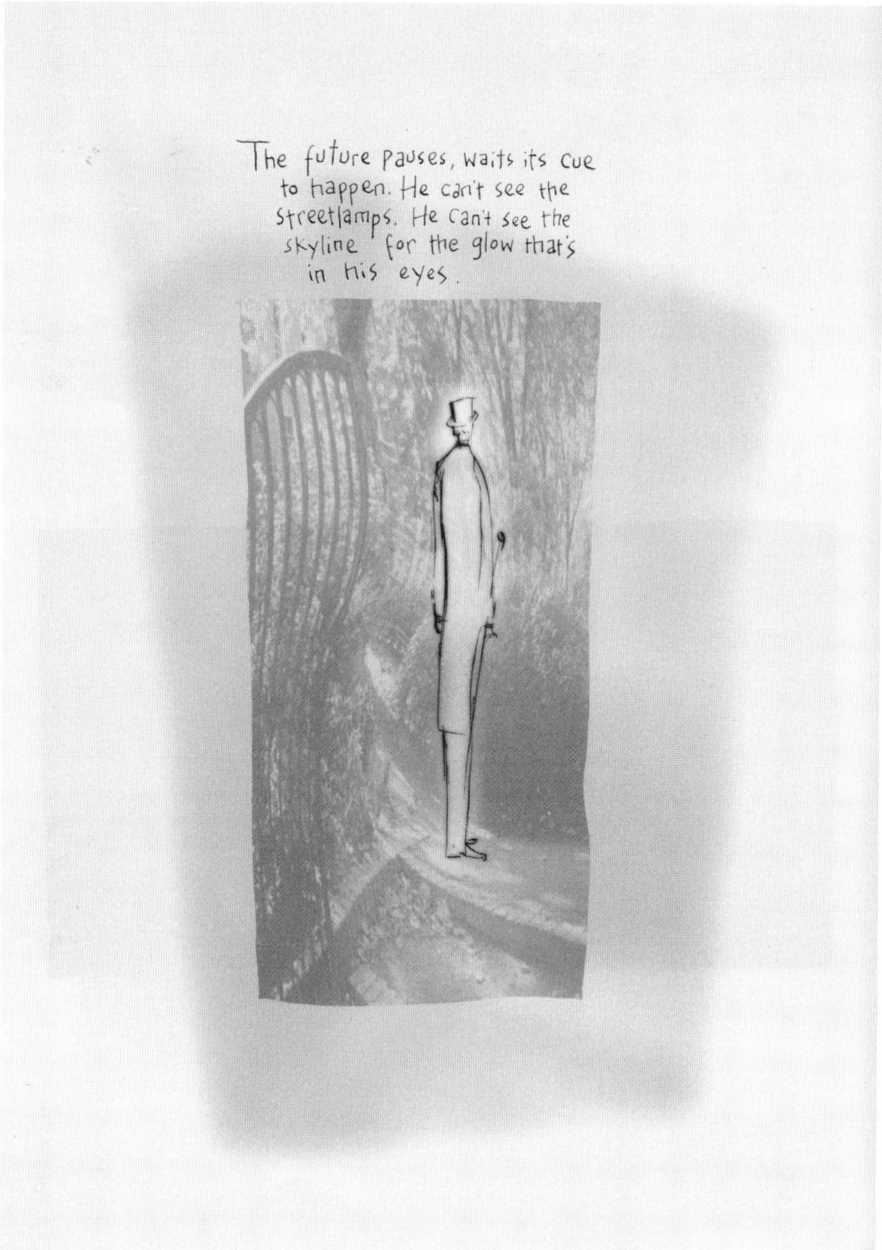

Figure 2.4 (left and right): From 'V Syon', Snakes and Ladders.

"...for the glow that's in his eyes."

Figure 4.1: "A Curious Woman", Warnauts, E. and G. Raives (1996): Lettres d'Outremer. Tournai, p.28.

Figure 4.2: "Natural beauty", Warnauts, E. and G. Raives (1996): Lettres d'Outremer. Tournai, p. 67.

Figure 4.3: "The timeless, exotic woman", WARNAUTS, E. and G. RAIVES (1996): Lettres d'Outremer. Tournai, p. 98.

Figure 4.4: "Leila and Landscape", LAPIERE, D. and J-P STASSEN. (1999): Le Bar du vieux Français (Edition intégrale). Paris, p. 53.

Figure 4.5: "Boxed In", LAPIÈRE, D. and J-P STASSEN. (1999): Le Bar du vieux Français (Edition intégrale). Paris, p. 116.

Figure 5.1: Carefully-observed mundane details, connecting the personal to the wider geopolitical landscape (SACCO 2001, 222).

Figure 5.2: Frames of war and conflict are presented as being constructed by multiple actors, as boundaries are performed and represented (DELISLE 2011, 40).

Figure 5.3: Vicarious experiences and empathy: staging torture to feel it in one's own body, while recognising the potential and unease of the voyeurism this implies, in the mundane setting of a home (Sacco 2001, 94).

Figure 5.4: Mundane daily life in exceptional circumstances: carefully-observed vignettes of how the wider geopolitical context contaminates the scale of individual and family life (DELISLE 2011, 49).

Figure 5.5: "I had never seen the wall of separation. I didn't think it was that big." The scale of human experience is confronted with the represented scale of the divided geopolitical landscape (DELISLE 2011, 37).

Figure 5.6: Le Roy 2010, 7. Picturing boundedness, dreaming escape.

Figure 5.7: Like Delisle, Le Roy uses a mise-en-abîme to show the moment of creation of images, either himself drawing what Mahmoud is drawing (Figure 5.6) or using one of Mahmoud's images (Figure 5.7) (Le Roy 2010, 44).

Figure 5.8: Mahmoud's depiction of dialogue, reduced to scribbles and dots, gives a surreal, nighmareish flavour to his drawings, providing direct depictions of the experience of violence, as lived in the first person (LE ROY 2010, 64).

Figure 5.9: Caged lives and bounded individual ("By the way, did you get back alright the other day?" –"Yes, fine, thanks.") (LE ROY 2010, 75).

Figure 6.1: Discussing Jugoslava on the trip back to Sarajevo (Copyright Joe Sacco, images courtesy of Drawn & Quarterly).

Figure 6.2: Contrasting Karadžić's and Rada's depictions in the narrative (Copyright Joe Sacco, images courtesy of Drawn & Quarterly).

Figure 7.1: Decaying landscapes (The Walking Dead #2).

Figure 7.2: A 'living' mass (AMC's The Walking Dead).

Figure 7.3: Entering Atlanta (The Walking Dead #2).

Figure 7.4: Cryptomimetic reveal (The Walking Dead #9).

Figure 7.5: Cryptomimetic reveal (The Walking Dead #17).

Figure 8.1: Panels taken from Avengers #500 (2004). © 2013 Marvel.

Figure 8.2: Whole page taken from House of M #5 (2005). © 2013 Marvel.

Figure 8.3: Whole page taken from Uncanny X-Men #463 (2005). © 2013 Marvel.

Figure 8.4a: Front cover of House of M #1 (2005) and panel from House of M #7 (2005).
© 2013 Marvel.

Figure 8.4b: Front cover of House of M #1 (2005) and panel from House of M #7 (2005).
© 2013 Marvel.

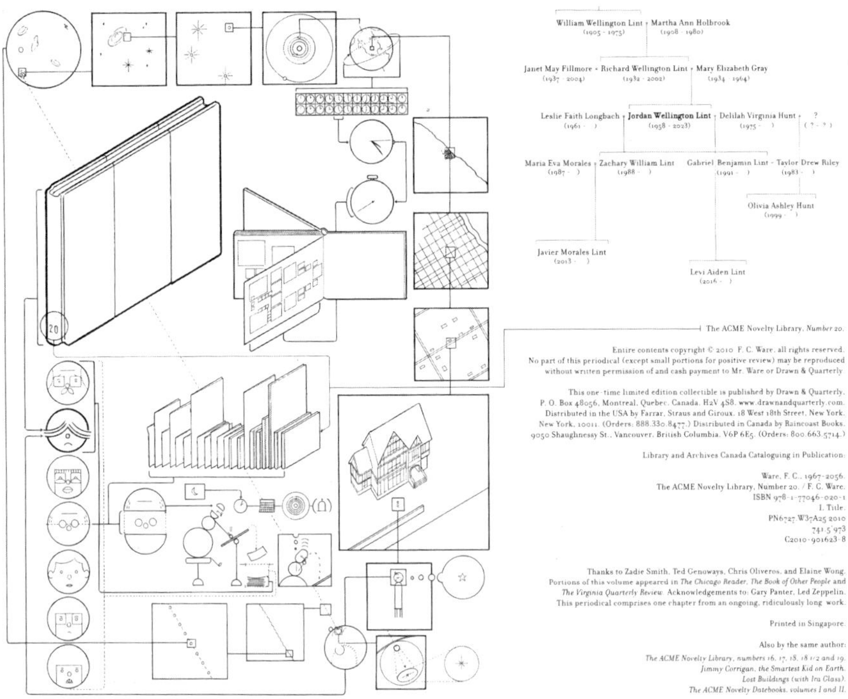

Figure 9.1: Front end-paper, The Acme Novelty Library Number 20. © F. C. WARE.

Figure 9.2: 'Quimby the Mouse' cartoon strip, Quimby the Mouse. © F. C. Ware.

Figure 9.3: 'Quimby the Mouse' cartoon strip, Quimby the Mouse. © F. C. Ware.

Figure 9.4: Rear end-paper, The Acme Final Report to Shareholders and Saturday Afternoon Rainy Day Fun Book. © F. C. Ware.

Figure 9.5: The Acme Final Report to Shareholders and Saturday Afternoon Rainy Day Fun Book. © F. C. Ware

Figure 9.6: The Acme Novelty Library Number 20. © F. C. WARE.

Figure 9.7: Jimmy Corrigan, the Smartest Kid on Earth. © F. C. WARE.

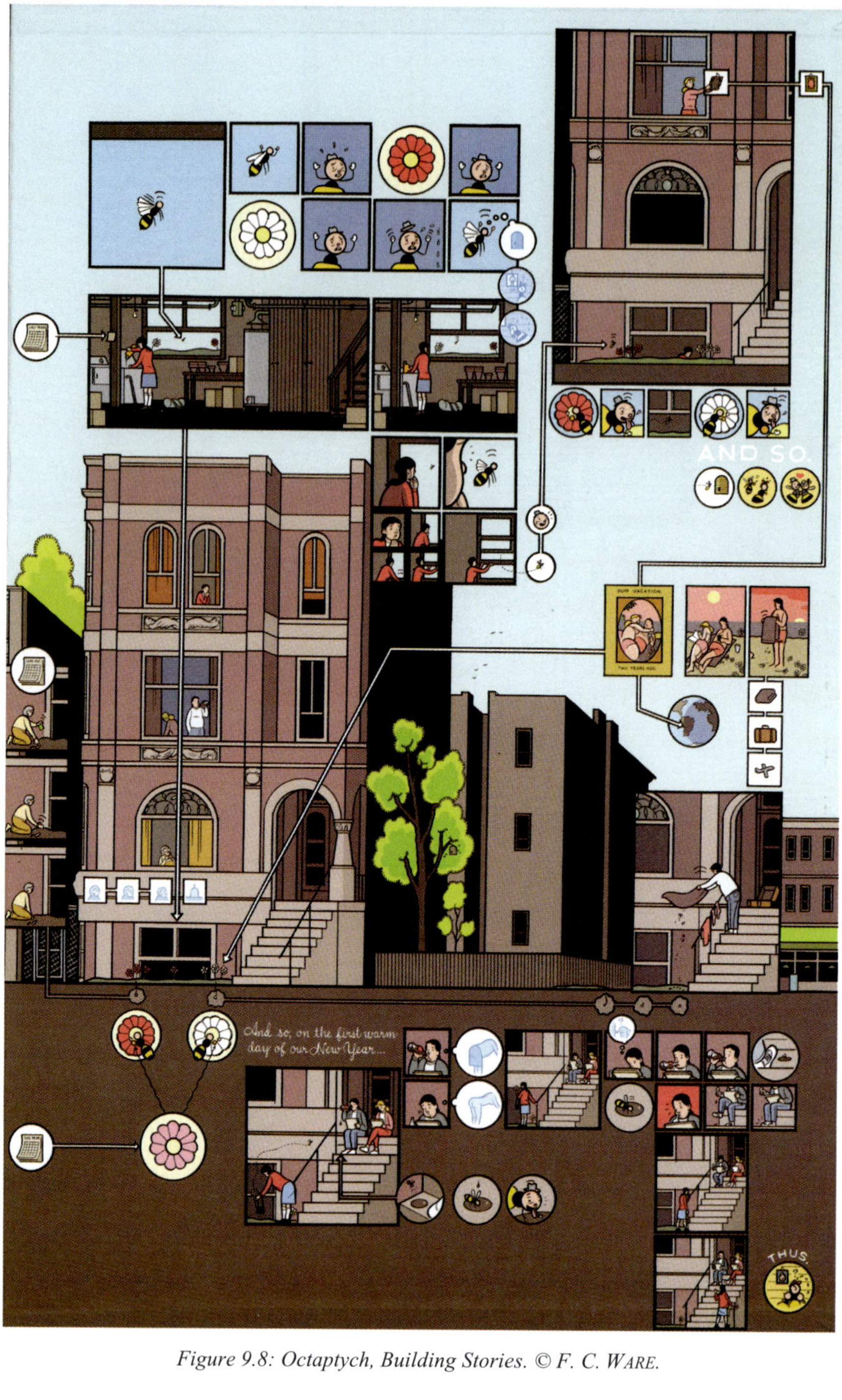

Figure 9.8: Octaptych, Building Stories. © F. C. WARE.

Figure 9.9: Building Stories. © F. C. Ware.

CONTRIBUTORS

Jason Dittmer is Reader in Human Geography at University College London. Dittmer's research interests are broadly within the realm of political geography, and more specifically on the role of the everyday in constituting geopolitical assemblages. He has published *Popular Culture, Geopolitics, and Identity* (2010, Rowman and Littlefield) and *Captain America and the Nationalist Superhero* (2013, Temple University Press) on the role of popular culture in geopolitics, and co-edited *Mapping the End Times: American Evangelical Geopolitics and Apocalyptic Visions* (2010, Ashgate) on the role of eschatology in shaping early 21st century geopolitics. His current project examines diplomacy as everyday flows of information and materials.

Marcus A. Doel is Professor of Human Geography at Swansea University, and Co-Director of the University's Centre for Urban Theory. He is the author of *Poststructuralist Geographies: The Diabolical Art of Spatial Science* (Edinburgh University Press, 1999), and co-editor of books including *Jean Baudrillard: Fatal Theories* (Routledge, 2009), *Moving Pictures/Stopping Places: Hotels and Motels on Film* (Lexington, 2009), and *The Consumption Reader* (Routledge, 2004). He has published extensively on poststructuralist social and spatial theory, modern and postmodern cities, and the history and philosophy of Geography.

Oliver Dunnett is a cultural, historical and political geographer whose research interests primarily focus on the ways in which cultures of outer space, science and technology are connected to questions of place, landscape and identity in the Western world. Oliver has developed these interests through examining geographies of popular culture, including the medium of comics and the genre of science fiction. He has published work on the geographies of 'British outer space' and issues of geopolitics and identity in the Tintin comics. He is currently researching geographies of light pollution in late-twentieth century Britain, and has recently taken up a position as lecturer in human geography at Queen's University Belfast.

Juliet J. Fall is an Anglo/Swiss political geographer working on questions of biosecurity, borders and national identity. She is interested in how discourses of nature are enrolled into politics, and has explored how references to the environment are subverted to underpin the construction of political spaces. Recent research has dealt with the transboundary planning, protected areas, invasive species, ideas of 'natural borders', and urban gardening. She also works on the spaces and politics of knowledge production in a science studies perspective, and on the history of geography in Anglophone and Francophone contexts. She is a Full Professor at the University of Geneva, in Switzerland. She serves on the scientific boards of a number of journals, including *Progress in Human Geography, Environment and Planning D: Society and Space, Geographical Journal, L'Espace Politique* and *Geographica Helvetica.*

Daniel Merlin Goodbrey is a senior lecturer in Narrative & Interaction Design at The University of Hertfordshire in England. A prolific and innovative comic creator, Goodbrey has gained international recognition as a leading expert in the field of experimental digital comics. His hypercomic work received the International Clickburg Webcomic Award in Holland in 2006 while his work in print was awarded with the Isotope Award for Excellence In Comics in San Francisco in 2005. His smartphone app, A Duck Has An Adventure, was shortlisted in the 2012 New Media Writing Prize.

Michael Goodrum received his PhD from the University of Essex in 2011, where he works in the Department of History. His research is in US cultural history, with a particular interest in the ways in which trauma, gender and race are represented in comic-books. He is co-editor of a forthcoming book on Joss Whedon's TV series *Firefly*, has been published in *Social History*, and has a chapter in *The Joker: Critical Essays on the Clown Prince of Crime*, due for publication in 2014. Current research focuses on the representation of political protest in superhero narratives.

Edward C. Holland is a human geographer with interests in a wide range of topics, including critical geopolitics, political violence, and religion. He has previously published on Joe Sacco's work in *Geopolitics* and is preparing a chapter for inclusion in a forthcoming, multidisciplinary volume on Sacco's comics. Holland currently holds a postdoctoral fellowship at the Havighurst Center for Russian and Post-Soviet Studies at Miami University. Formerly, he was a Title VIII Research Scholar affiliated with the Kennan Institute at the Woodrow Wilson International Center for Scholars.

Shaun Huston is an associate professor of geography and film studies at Western Oregon University. His documentary, Comic Book City, Portland, USA (2012), can be viewed on Vimeo and at the Internet Archive. He writes regularly on comics, film, and popular culture for PopMatters.

Catriona MacLeod is lecturer at the University of London Institute in Paris. Her research focuses on gender and postcolonialism in twentieth century bande dessinée. Her future research projects include studies of the carnival of French Guyana, particularly the mysterious figure of the 'Touloulou' and the mélange of caricature and orientalism in the paintings of Gustave-Henri Jossot.

Julia Round is senior lecturer in the Media School at Bournemouth University, UK, and co-edits the academic journal *Studies in Comics* (Intellect Books). Her research and teaching interests include gothic, comics, adaptation and children's literature. She is currently completing her monograph *Gothic in Comics and Graphic Novels: A Critical Approach* (McFarland, 2014) and co-editing the collection *Real Lives, Celebrity Stories* (Continuum 2014). For further details please visit www.juliaround.com.

Tony Venezia recently completed a PhD on comics writer Alan Moore and the historical imagination in his work at Birkbeck, University of London. His research interests are in British and American comics and genre writing. He is a teaching fellow and visiting lecturer teaching English and film at Birkbeck and London South Bank University, and has published articles and reviews in a number of academic journals including *Peer English, Radical Philosophy, New Formations,* and *Visual Culture in Britain.* He has contributed to the forthcoming collection of essays *Alan Moore and the Gothic Tradition,* and is on the editorial board of the journal *Studies in Comics.* He is co-convenor of the Contemporary Fiction Seminar series at the Institute of English Studies. He is currently co-editing essay collections on the science fiction/fantasy writer China Miéville and on twenty-first century British fiction. Tony established the annual *Transitions* symposium at Birkbeck, the only annual London-based comics symposium and currently on its fourth year.

Franz Steiner Verlag

Place, Television, and the Real Orange County
By Ann Fletchall, Chris Lukinbeal and Kevin McHugh

2012.
144 pages with 84 illustrations.
Softcover.
ISBN 978-3-515-10118-9

Place, Television, and the Real Orange County

By Ann Fletchall, Chris Lukinbeal and Kevin McHugh

Media Geography at Mainz – Volume 2

This case study is an exploration of televisual place, of Orange County, California, and three popular U.S. shows set therein: *The OC, Laguna Beach: The Real Orange County*, and *The Real Housewives of Orange County*. Place is a meaningful experience in the world, and it is made through a unique intersection of social processes. It is also an amalgam of memory, emotion, and imagination. Places of the media fit this description.

Mediated places are an inextricable part of our daily lives, and directly engage the processes of place-making by affecting our perception, senses, and subjectivity. These three Orange County based series are used to demonstrate how production techniques contribute to the place-making process and how this process continues and culminates with audience engagement. The use of landscape images, the concept of emotional realism, and reality TV's claim to the real are explored for their role in the televisual place-making process. Audience surveys and the phenomenon of TV-induced tourism demonstrate the importance of televisual places to viewers. This book proves that mediated places matter.

Contents

Franz Steiner Verlag
Birkenwaldstr. 44 · D – 70191 Stuttgart
Telefon: 0711 / 2582 – 0 · Fax: 0711 / 2582 – 390
E-Mail: service@steiner-verlag.de
Internet: www.steiner-verlag.de